Springer Tracts in Modern Physics
Volume 122

Editor: G. Höhler
Associate Editor: E. A. Niekisch

W0232089

Editorial Board:
S. Flügge H. Haken J. Hamilton
W. Paul J. Treusch

Springer Tracts in Modern Physics

Volumes 90–106 are listed on the back inside cover

Particle Induced Electron Emission I

With Contributions by
M. Rösler, W. Brauer
and
J. Devooght, J.-C. Dehaes, A. Dubus,
M. Cailler, J.-P. Ganachaud

With 64 Figures

Springer-Verlag Berlin Heidelberg GmbH

Dr. Max Rösler
Zentralinstitut für Elektronenphysik, Hausvogteiplatz 5–7,
O-1086 Berlin, Fed. Rep. of Germany

Professor Dr. Wolfram Brauer
Uhlandstr. 35, O-1110 Berlin, Fed Rep. of Germany

Professor Dr. Jacques Devooght
Professor Dr. Jean-Claude Dehaes
Professor Dr. Alain Dubus
Université Libre de Bruxelles, Service de Métrologie Nucléaire (CP 165)
50, av. F. D. Roosevelt, B-1050 Brussels, Belgium

Dr. Michel Cailler
Université de Nantes, ISITEM, Laboratoire de Sciences des Surfaces et Interfaces
en Mécanique, La Chantrerie (CP 3023), F-44087 Nantes CEDEX 03, France

Professor Dr. Jean-Pierre Ganachaud
Université de Nantes, Faculté de Sciences et des Techniques, Laboratoire de Physique
du Solide Théorique, 2, rue de la Houssinière, F-44072 Nantes CEDEX 03, France

---

Manuscripts for publication should be addressed to:
Gerhard Höhler
Institut für Theoretische Kernphysik der Universität Karlsruhe, Postfach 69 80,
W-7500 Karlsruhe 1, Fed. Rep. of Germany

*Proofs and all correspondence concerning papers in the process of publication
should be addressed to:*
Ernst A. Niekisch
Haubourdinstraße 6, W-5170 Jülich1, Fed. Rep. of Germany

ISBN 978-3-662-14999-7   ISBN 978-3-540-48723-4 (eBook)
DOI 10.1007/978-3-540-48723-4

Preface

This book is devoted to low energy electron emission from solid surfaces bombarded by electrons or ions. The most recent theoretical models of electron emission induced by low energy electrons ($E \leq 10\text{keV}$) and by medium energy protons ($10\text{keV} \leq E \leq 1\text{MeV}$) are reviewed.

Although the practical application of electron emission is not considered, the subject discussed in this book is of great importance in processes such as plasma-wall interactions, scanning electron microscopy and particle detection.

When charged particles penetrate a solid target, electrons of a few eV are emitted as a consequence of the excitations of target electrons by the incident particle. If the excitation energy comes from the kinetic energy of the incident particle, this emission phenomenon is called kinetic secondary electron emission (incident electrons) or kinetic ion-induced electron emission (incident ions).

The modelling of electron emission involves a detailed description of the most important interactions that a charged particle can undergo in a solid and of the transport process of the incident and excited particles. Both aspects are emphasized in this volume.

The contribution by Rösler and Brauer gives a detailed theory for nearly-free-electron metals. It includes the definition of the basic quantities, a description of the basic concepts concerning the emission phenomena, a discussion of the scattering functions and mean free paths of the electrons and the numerical treatment of the transport of the excited electrons towards the surface.

The contribution by Devooght et al. presents complementary aspects on the interaction of charged particles in solids, including a discussion on the extension to non-free-electron solids, and a description of the electron transport models based upon the Monte Carlo method and upon the Boltzmann equation. Various methods to solve the Boltzmann equation are given.

Both contributions give results for aluminum. In particular, the energy and angular distribution of the emitted electrons and the electron yield are discussed. Other aspects are also considered: the influence of the primary particle transport, the role of the elastic scattering, statistical aspects of the emission process, etc.

Berlin and
Brussels, March 1991

M. Rösler and W. Brauer
J. Devooght, J.-C. Dehaes, A. Dubus
M. Cailler, and J.-P. Ganachaud

Contents

Theory of Electron Emission from Nearly-Free-Electron Metals by Proton and Electron Bombardment

M. Rösler and W. Brauer

With 42 Figures

1. Introduction

In the processes of ion-induced electron emission (IIEE) and secondary electron emission (SEE) electrons are emitted from the surface of a solid as a result of its bombardment by ions and primary electrons (PE), respectively. Thus, electron emission is a consequence of the inelastic interaction between the incident particles and the solid state electrons. Both phenomena were discovered at the beginning of this century (Thomson 1904; Rutherford 1905; Austin and Starke 1902).

The particle-induced electron emission is utilized for particle detection. In connection with problems of plasma-wall interactions in thermonuclear fusion reactors the ejection of electrons is of great importance. Furthermore, electron emission complicates all measurements of particle currents in irradiation experiments. However, the most important application of the emission phenomenon is scanning electron microscopy (SEM). In this case visible images of small sample areas can be obtained by variation in the yield of emitted electrons. In the present article aspects of practical application of the particle-induced electron emission will not be considered. Rather, the aim is to formulate a common microscopic description of the basic processes determining both emission phenomena.

In a recent review paper of *Schou* (1987) concerning electron emission from solids by electron and proton bombardment, an extensive list of reviews on SEE and IIEE can be found. Besides the very recent paper by *Bindi, Lanteri,* and *Rostaing* (1987) which emphasizes the microscopic description of the underlying basic processes, no comprehensive representation of the theory of SEE has been published in the last two decades. Most of the papers having appeared in the last few years are devoted to aspects connected with SEM. For IIEE recent reviews have been presented by *Sigmund* and *Tougaard* (1981) and *Hasselkamp* (1985). In both reviews general aspects of SEE are included as well. Recent results in the field of kinetic IIEE from solids under bombardment by energetic ions have been summarized by *Hasselkamp* (1988).

From the large number of existing experimental results we single out those data which are appropriate to support theoretical considerations and are important for a comparison with our quantitative results.

The present contribution is devoted to the theory of *kinetic* IIEE and SEE. In both cases electrons are excited within the solid by direct transfer of kinetic

1

energy from the impinging particle. At low ion energies IIEE proceeds in front of the solid surface. The theory of this so-called potential emission will be comprehensively discussed in a later volume of this series by Varga.

A common theoretical description of kinetic IIEE and SEE is possible because the transport of excited electrons within the solid as well as the escape of secondary electrons (SE) are determined only by the properties of the target. But also the excitation process reveals common features because the basic interaction between the charged primary particle and the solid state electrons is the screened Coulomb interaction in both cases.

It is beyond the scope of this book to formulate a general microscopic theory of particle-induced electron emission including all types of solids and experimental conditions. There are a number of restrictions. First of all, we formulate a theory for nearly-free-electron (NFE) metals. Only in this case is a simple description of electronic structure using the model potential formalism possible with sufficient accuracy. It should be noted, however, that numerous aspects of our theoretical considerations can be applied to the description of emission phenomena for other types of target materials. In the case of kinetic IIEE we will concentrate our consideration on a theory of impact by ions with medium to high energies. Proton impact will be discussed in particular detail, because in this case complications due to projectile electrons can be avoided.

In order to simplify the mathematical description of the emission problem, normal incidence of the primary beam will be considered. Most of the experiments were carried out on polycrystalline metals. Therefore, a restriction to such targets seems to be reasonable. With respect to a unified description of kinetic IIEE and SEE, the stopping of the primary particle in the region below the surface, which is important for the emission phenomenon, should be neglected. For kinetic IIEE this assumption is fulfilled at medium and high ion energies. In the case of SEE this assumption requires the primary energies to be larger than about 1 keV.

In Chap. 2 the basic quantities (energy-angular distribution, energy distribution, electron yield) are defined. In order to compare to our theoretical results we present a selection of experimental data for these quantities. Chapter 3 is devoted to the basic concepts concerning the emission phenomena. Chapter 4 contains a simple description of the escape process of the liberated electrons. Starting from a Boltzmann equation formulation the transport of inner excited electrons is treated in Chap. 5. The scattering functions and mean free paths which govern the slowing-down and diffusion of excited electrons towards the surface of the target are discussed in Chap. 6. The different excitation rates appearing in NFE metals are evaluated in Chap. 7. In Chap. 8 the numerical treatment of the Boltzmann equation is discussed. Results for aluminum are given in Chap. 9. Starting from these results the effect of elastic scattering on the emissive properties is discussed in Chap. 10. Finally, Chap. 11 is devoted to miscellaneous problems in connection with particle-induced electron emission. A historical survey is given in the Appendix.

2. Basic Quantities in IIEE and SEE

In this chapter the basic quantities for the description of emission phenomena are introduced. Further on, some experimental results for aluminum, especially in the case of IIEE, are presented. These data are important for a comparison with our theoretical results as well as for the development of basic concepts.

The number of electrons with energy E emitted per second from $1 \, cm^2$ of the surface in the direction Ω, i.e., the energy and angle dependent current density $j(E, \Omega)$, is the basic quantity for the description of emission phenomena neglecting spin. $j(E, \Omega)$ is related to the unit current of particles impinging on the surface with energy E_0. The maximum information about the emission process can be obtained by measuring this quantity. However, the actual experimental quantities are the energy distribution

$$ j(E) = \int j(E, \Omega) \, d\Omega \,, \tag{2.1} $$

the angular distribution

$$ j(\Omega) = \int j(E, \Omega) \, dE \,, \tag{2.2} $$

and the electron yield.

In the case of IIEE the electron yield, usually denoted by γ, is simply given by

$$ \gamma = \int j(E) \, dE \,. \tag{2.3} $$

In general, γ is composed of the contributions of impinging ions and recoil atoms. The latter contribution is important for heavy ions at low E_0. However, for protons the contribution of recoil atoms to the electron emission can be neglected at the proton energies considered here (Sigmund and Tougaard 1981).

In the case of SEE the spectrum of outgoing electrons is complicated by the backscattering of PE. Figure 2.1 shows schematically the energy distribution $j(E)$ of electrons released by PE. The emitted electrons can have all energies up to the primary energy E_0. The outgoing electrons are conventionally divided into different groups: The true SE with energies below $50 \, eV$ and the inelastically backscattered electrons as well as the elastically reflected PE with energies above $50 \, eV$ up to E_0. Superimposed on the energy distribution there are often some peaks which are produced by Auger processes accompanying the ionization of inner shell electrons.

According to the subdivision of the outgoing electrons, the total secondary yield σ can be written as

$$ \sigma = \int_0^{E_0} j(E) \, dE = \int_0^{50} j(E) \, dE + \int_{50}^{E_0} j(E) \, dE \equiv \delta + \eta \,. \tag{2.4} $$

3

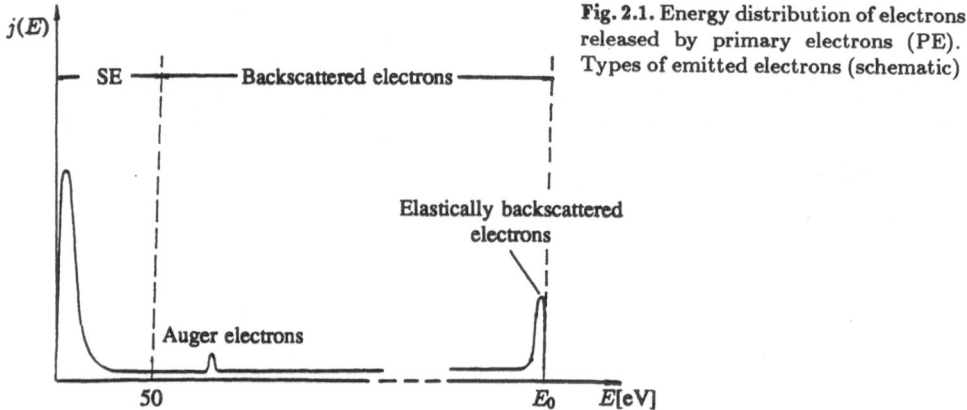

Fig. 2.1. Energy distribution of electrons released by primary electrons (PE). Types of emitted electrons (schematic)

δ represents the true secondaries and η the backscattered electrons. The yield of true SE is given by the contribution of incident (δ_p) and backscattered ($\eta\delta_r$) PE,

$$\delta = \delta_p + \eta\delta_r .\qquad\qquad(2.5)$$

Sometimes one expresses the contribution δ_r by the efficiency β of backscattered electrons according to

$$\delta_r = \beta\delta_p .\qquad\qquad(2.6)$$

The secondary electron yield δ and the coefficient of backscattered electrons η can be measured using standard set-ups. The current collected from the target is measured with a hemispherical grid biased at $-50\,\mathrm{eV}$ and $50\,\mathrm{eV}$. In this way, one obtains the backscattering coefficient η and the total yield σ, respectively. δ is then determined from the difference of these quantities. The present work is concerned only with δ_p. Measurements of the coefficients δ_p and δ_r must be performed separately. From experiments using thin films (self-supporting or deposited on a substrate of a different material) of various thicknesses δ is obtained as a function of η and, therefore, δ_r can be evaluated from (2.5).

There are two different lengths essential for the description of both emission phenomena. The first is the escape depth L of SE which is of the order of 5 nm in metals. The second is the range of the incident charged particles R depending on their energy E_0. If the range of the incoming particle is large compared with the escape depth, one can approximate the trajectories of the primary particles throughout this depth by straight lines. This assumption has simplifying consequences for the theoretical description of emission phenomena because apart from the surface of the specimen there is no further spatial dependence in the problem.

In the case of SEE the condition $R(E_0) \gg L$ means that the primary energy E_0 is restricted to values larger than about 1 keV corresponding to a range of the PE of $R \approx 40$ nm (Fitting 1974). In the case of IIEE the range of ions (here

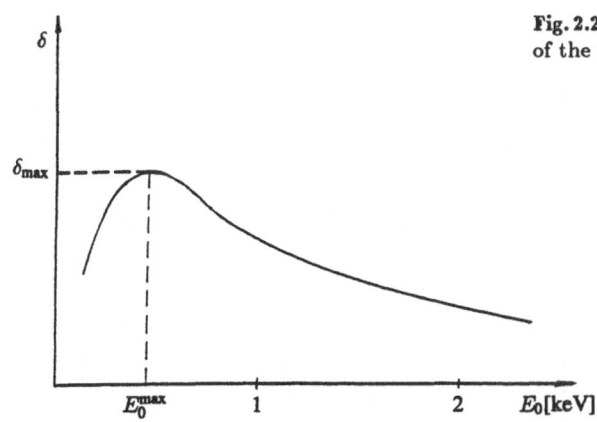

Fig. 2.2. SEE. Electron yield as a function of the primary energy (schematic)

H^+) in metals is roughly two orders of magnitude larger than the escape depth for projectile energies above 20 keV considered in this paper.

The general shape of the primary energy dependence of the electron yield δ is shown in Fig. 2.2. For all materials δ rises with increasing E_0, reaches a maximum and finally decreases for higher primary energies. For metals the maximum is in the 100 eV range (for aluminum we have $E_0^{\max} \approx 300\,\text{eV}$). The mentioned restriction to primary energies larger than about 1 keV implies that our theoretical considerations are valid only for primary energies beyond the value $E_0^{\max}$.

In Fig. 2.3 experimental results for proton impact on Al obtained by different groups (Baragiola, Alonso, and Oliva Florio 1979; Hasselkamp et al. 1981; Svensson and Holmén 1981; Koyama, Shikata, and Sakairi 1981) are collected. The general behavior of yield vs. ion energy should be the same for other ion-target combinations. However, so far the maximum in the yield curves has been obtained only for light ions (H^+, H_2^+, D^+, and He^+).

The reasons for the appearance of yield maxima in SEE and IIEE are different (Rösler and Brauer 1984). In the latter our theoretical considerations

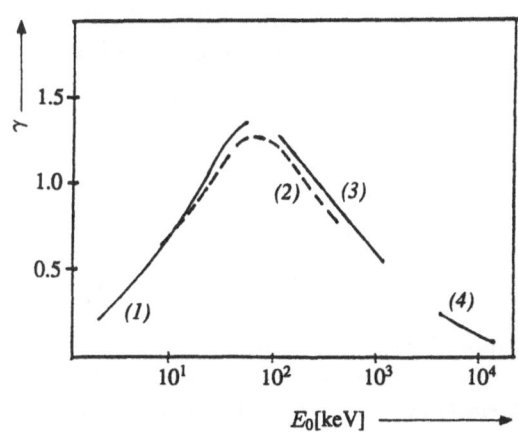

Fig. 2.3. IIEE. Electron yield γ as a function of the ion energy for proton impact on aluminum. Experimental values: (1) Baragiola, Alonso, and Oliva Florio (1979); (2) Svensson and Holmén (1981); (3) Hasselkamp et al. (1981); (4) Koyama, Shikata, and Sakairi (1981)

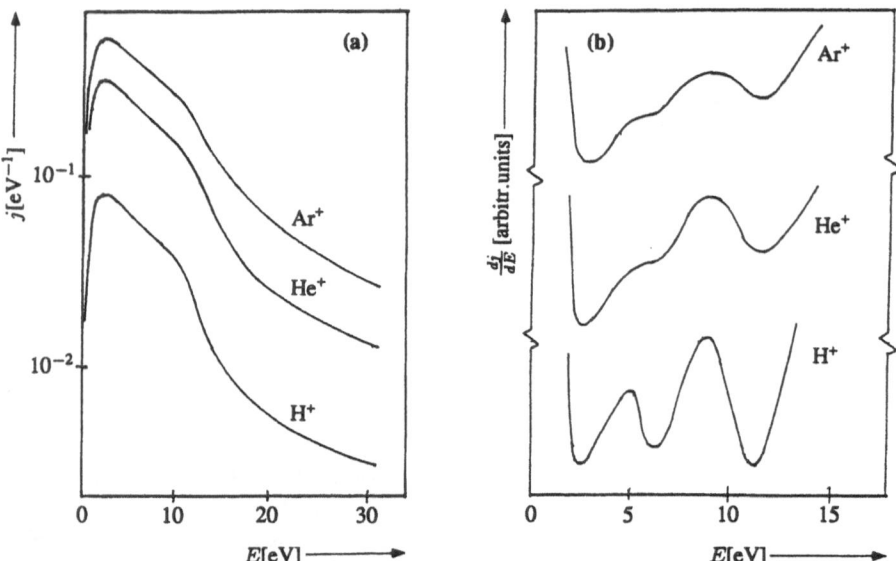

Fig. 2.4. Energy distribution of ejected electrons (a) and the corresponding derivative spectra (b) from Al for different projectile ions (Hasselkamp and Scharmann 1982). $E_0 = 500\,\mathrm{keV}$

should be valid in the whole energy range of impinging protons including the yield maximum.

Figure 2.4a shows the energy distributions $j(E)$ for different projectile ions (H^+, He^+, Ar^+) at $E_0 = 500\,\mathrm{keV}$ measured by *Hasselkamp* and *Scharmann* (1982). In the energy distribution a main maximum at $2\,\mathrm{eV}$ and two weak shoulders at $11.5\,\mathrm{eV}$ and $6\,\mathrm{eV}$ are observed for all types of projectiles. These structures are clearly seen in the corresponding differential spectra (Fig. 2.4b). The complete agreement with the corresponding curves for SEE in Al (Everhart et al. 1976) provides a distinct experimental clue for developing a common theory of both SEE and IIEE.

Finally, we mention that the energy-integrated angular distribution observed for SEE in Al (Oppel and Jahrreis 1972) is cosine-like. This should be expected also for the angular distribution of emerging electrons at proton impact. Up to now, no measurements of this angular distribution have been published.

3. Basic Features

In this chapter we discuss the basic principles and review different recent models for the description of emission phenomena.

Kinetic electron emission by bombardment with fast charged particles is usually described by three stages. First, internal electrons are excited by the incident particle. Second, these excited electrons interact with the medium leading to the slowing-down of these electrons as well as to their migration to the surface. Last, the electrons which reach the surface may escape through the potential barrier. In practice, all treatments of particle-induced electron emission use this three-step model.

We emphasize that both emission phenomena are, in fact, governed by bulk processes. Nevertheless, the surface of the target plays a leading part in particle-induced emission processes. First, it is the potential barrier W at the surface which influences the escape of electrons into the vacuum. This will be made clear by measurements of the energy distribution of ejected electrons by proton impact on Al using targets with clean surfaces as well as surfaces covered with an oxide layer (Pillon, Roptin, and Cailler 1976; Hasselkamp 1985). This surface barrier effect will be taken into account in the simple description of the escape process (stage 3) given in the next chapter.

In addition to this fundamental surface effect there are elementary processes by which electrons are excited directly at the surface. As can be seen in Fig. 2.4b there is a distinct minimum in the derivative of the energy distribution of Al for IIEE at 6 eV. This structure, which appears also in the case of SEE (Everhart et al. 1976), can be attributed to the decay of surface plasmons excited by the incident charged particle. However, calculations have shown (Chung and Everhart 1977); Ganachaud and Cailler 1979) that the number of electrons excited by this elementary process is small compared with the total number of electrons excited via different bulk processes. Therefore, for the electron yield surface plasmon effects are negligible.

Recently developed models which are used for the description of particle-induced electron emission can be classified into three different categories. First, we should mention *Schou*'s model (Schou 1980; 1987) which is based on the analogy between electron emission and sputtering. This model uses the energy-deposition law in the target and macroscopic properties (stopping power) as input quantities. Second, are the numerous Monte-Carlo calculations of SEE (Fitting and Reinhart 1985; Ganachaud and Cailler 1979; Koshikawa and Shimizu 1974; Shimizu and Ichimura 1983; Valkealathi and Nieminen 1984; Lyo and Joy 1988). In these papers different approximations concerning the microscopic scattering cross sections of electrons in the target are used. Third, are the models based on the solution of the Boltzmann transport equation. In order to determine the number of electrons leaving the target at the surface there are different approaches to the treatment of boundary and escape conditions. A general formulation of the transport equation with suitable boundary conditions was given by *Puff* (1964). In such an approach the distinction between the last two stages of the three-step model mentioned above is not necessary.

The Boltzmann transport equation with the boundary conditions formulated by Puff was solved numerically by *Bindi. Lanteri*, and *Rostaing* (1980a) and included the spatial dependence of the excitation rate as a consequence of scattering and energy loss of PE. Except for the excitation of core electrons, all elementary processes essential for the description of SEE in NFE metals are taken into consideration. However, because of the large numerical effort required for solving the integro-differential equation in question it is necessary to make some simplifying assumptions concerning the different excitation processes and scattering cross sections.

Two recent transport models for particle-induced electron emission developed by *Devooght*, *Dubus*, and *Dehaes* (1987) and *Dubus*, *Devooght*, and *Dehaes* (1986) consider also the semi-infinite character of the target. As a consequence, the escape process (step 3) is automatically included. Discussions of both models have been recently published (Dubus, Devooght, and Dehaes 1989) and are also conducted here.

In the present work we will employ the infinite-medium slowing-down model which uses a simple description of the escape process given in the next chapter. Special emphasis is directed to a consistent treatment of different interaction processes concerning excitation as well as slowing-down (Rösler and Brauer 1988). In all previous works the basic quantities (source functions, mean free paths, differential scattering cross sections) are calculated using different approximations for the same underlying interaction processes. In particular, for analytical or simple numerical solution of the Boltzmann equation special approximations for the scattering cross sections are customary, such as the widely used simple electron-electron scattering function proposed by *Wolff* (1954).

It is important to note that the slowing-down of excited electrons is determined in fact by electron-electron interaction. Therefore, it is necessary to calculate the corresponding scattering cross sections with high accuracy taking into account the dynamical screening of the Coulomb interaction (Rösler and Brauer 1988). However, the solution of the Boltzmann transport equation using source functions, mean free paths, and differential scattering cross sections evaluated on the same level including dynamical screening needs considerable numerical effort. Such a treatment improves the results of SEE (Rösler and Brauer 1981a; 1981b) and proton-induced electron emission (Rösler and Brauer 1984) in Al. In these papers the electron-electron scattering cross sections are calculated on the basis of the static Thomas-Fermi approximation of the screening function.

4. Escape of Secondary Electrons

In this chapter a simple description of the escape process is given which is applicable to NFE metals due to their close connection with the free-electron-gas model.

In order to describe the escape process we use the standard model of a planar surface barrier and free electrons inside the metal. Such a simple model should be valid if the diameter of the emitting area is large compared with the lattice constant. In this case the realistic potential can be replaced by a suitable constant mean value. The surface barrier of height W is determined in metals by the Fermi energy $E_F = \hbar^2 k_F^2/2m$ and the work function Φ, i.e., $W = E_F + \Phi$. For Al we use $E_F = 11.6\,\mathrm{eV}$ ($k_F = 1.75 \cdot 10^8\,\mathrm{cm}^{-1}$) and $\Phi = 4.3\,\mathrm{eV}$, respectively.

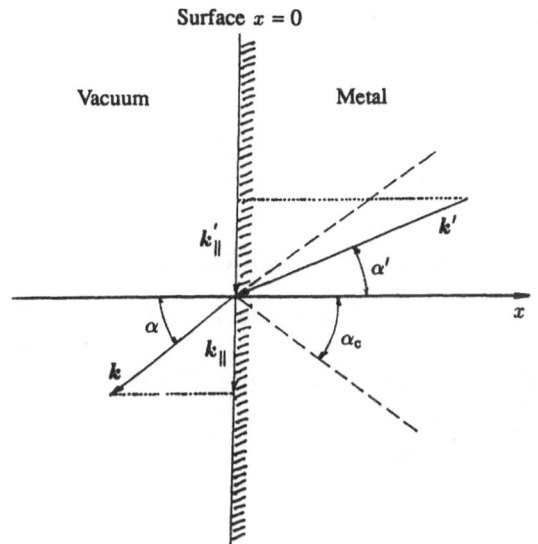

Surface $x = 0$

Vacuum

Metal

$k'_{\parallel}$

k'

α'

α

$k_{\parallel}$

α_c

k

x

Fig. 4.1. Momentum diagram for the escape process. k' and k are the momenta of the electron at both sides of the surface. The directions of these momenta (α', α) are measured with respect to the outer normal of the surface. α_c is the aperture of the escape cone. $k'_{\parallel}$ and $k_{\parallel}$ are the components of k' resp. k parallel to the surface

The conservation laws for energy $E = \hbar^2 k^2/2m = mv^2/2$ and parallel momentum connecting inner (E', Ω') and outer (E, Ω) variables are given by

$$E' = E + W$$
$$v'_{\parallel} = v(E')\sin\alpha' = v_{\parallel} = v(E)\sin\alpha \qquad (4.1)$$
$$\varphi' = \varphi ,$$

where Ω is the direction of the momentum of the electron

$$\Omega = (-\cos\alpha, \ \sin\alpha\cos\varphi, \ \sin\alpha\sin\varphi) \qquad (4.2)$$

measured with respect to the outer normal of the surface (Fig. 4.1); $v = v(E)$ is the velocity of the electron.

From (4.1) we obtain the following conditions necessary for the escape of an electron (E', Ω'):

$$E' > W$$
$$\cos\alpha' > \cos\alpha_c = \sqrt{\frac{W}{E'}} , \qquad (4.3)$$

where α_c is the maximum emission angle for which the normal component of momentum is sufficient for the electron to surmount the surface barrier. α_c defines the so-called escape cone depicted in Fig. 4.1. According to (4.1) α and α' are connected by the relation

$$\cos\alpha = \sqrt{\frac{E'\cos^2\alpha' - W}{E' - W}} . \qquad (4.4)$$

Because of particle conservation during the escape process we have on both sides of the surface $x = 0$

$$j(E, \Omega) \, dE \, d\Omega = j_i(E', \Omega') \, dE' d\Omega' \, . \tag{4.5}$$

The current density of inner excited SE at the surface $j_i(E', \Omega') = j_i \, (x = 0; E', \Omega')$ is related to the normalized electron density at the surface $N(E', \Omega')$ by

$$j_i(E', \Omega') = v(E') \cos \alpha' \, \Theta(E' - W) \, \Theta(\cos \alpha' - \cos \alpha_c) N(E', \Omega') \, , \tag{4.6}$$

where $\Theta(x)$ denotes the unit step function. $N(E', \Omega')$ is equal to the number of excited electrons in $1 \, \mathrm{cm}^3$ for unit primary current. Therefore, using (4.5, 1, 6) the current density $j(E, \Omega)$ measurable in free space can be written as

$$j(E, \Omega) = v(E') \left(1 - \frac{W}{E'} \right) \sqrt{\frac{E' \cos^2 \alpha' - W}{E' - W}}$$
$$\times \, \Theta(E' - W) \, \Theta(\cos \alpha' - \cos \alpha_c) N(E', \Omega') \, . \tag{4.7}$$

Some general statements can be obtained from (4.7) concerning the energy and angular behavior of the outer current density. In the limit $E \to 0$ it can be seen from the factor $1 - W/E'$ in (4.7) that $j(E, \Omega)$ goes to zero. Therefore, in this limiting case the outer energy distribution of electrons $j(E)$ will always start from the origin of the $E - j(E)$ coordinate system, as a direct consequence of the potential barrier at the surface. The experimental results confirm this statement. For $E \to 0$ a further result can be inferred from (4.7): $j(E, \Omega) \sim \cos \alpha$. This cosine law should be valid for very slow SE independent of the particular shape of $N(E', \Omega')$. If there is an isotropic distribution of inner SE at least within the escape cone, we obtain an exact cosine law for the angular distribution of outer SE at all energies.

5. Transport Theory of Excited Electrons

In this chapter the general form of Boltzmann's transport equation is given. This equation allows to determine the density of inner excited electrons in terms of excitation functions, mean free paths, and scattering functions. Starting from a dielectric formalism a unified description of these microscopic quantities is formulated.

5.1 Transition Probabilities. Dielectric Function.
Discussion of Different Elementary Processes

The different microscopic quantities governing the excitation of solid state electrons by the primary beam and the transport of inner excited SE are directly related via screened Coulomb interaction to the complex wave number and frequency dependent dielectric function $\varepsilon(q, \omega) = \varepsilon_1(q, \omega) + i\varepsilon_2(q, \omega)$ of the solid.

The general expression for the transition probability (neglecting exchange) between Bloch states for two interacting point charges screened by $\varepsilon(q,\omega)$ can be written as

$$W_{k_1 k_2 \to k'_1 k'_2} = \frac{32\pi^3 e^4}{\hbar} \frac{1}{\Omega} \sum_q \frac{1}{q^4 |\varepsilon(q, E_{k_1} - E_{k'_1})|^2} \, | < k'_1 |e^{iqr}|k_1 > |^2$$

$$\times | < k'_2 |e^{-iqr}|k_2 > |^2 \delta(E_{k'_1} + E_{k'_2} - E_{k_1} - E_{k_2}) , \qquad (5.1)$$

where Ω is the normalization volume. In the description of the transport process all particle states in (5.1) are related to electrons. The summation over the initial state $k_2(< k_F)$ (including the spin summation) and one of the final states results in the quantities

$$W_{k_1 \to k'_1} = \sum_{\substack{k_2(<k_F) \\ k'_2(>k_F)}} W_{k_1 k_2 \to k'_1 k'_2} \qquad (5.2)$$

and

$$W^s_{k_1 k'_2} = \sum_{\substack{k_2(<k_F) \\ k'_1(>k_F)}} W_{k_1 k_2 \to k'_1 k'_2} . \qquad (5.3)$$

The microscopic processes underlying the transition probability (5.2) and the excitation probability (5.3) are shown schematically in Fig. 5.1. (Only the interaction of the excited electron with conduction electrons is represented.)

The total transition probability $W(k', k)$ from state k to k' due to interaction with the system of solid state electrons consists of different contributions from single particle processes with conduction electrons (e) and core electrons (c) as well as plasmon processes (p) related to the system of conduction electrons:

$$W(k', k) = W_e(k', k) + W_c(k', k) + W_p(k', k) . \qquad (5.4)$$

The contributions from single particle processes (e and c) are given by (5.2)

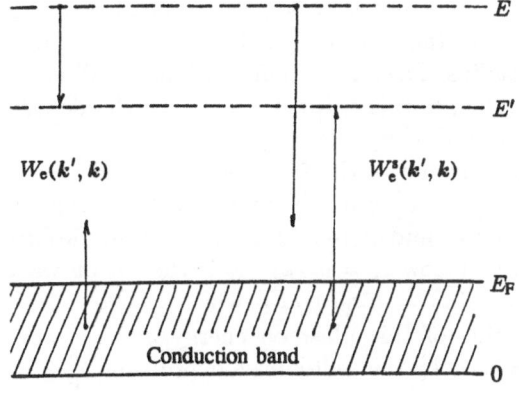

Fig. 5.1. The different electron-electron scattering processes for excited electrons with participation of the conduction band (schematic). See text

11

using the corresponding transition matrix elements in (5.1). With the help of the energy loss function $-\mathrm{Im}(1/\varepsilon)$ (Pines 1963) all contributions in (5.4) can be summarized in the representation

$$W(\mathbf{k}', \mathbf{k}) = -\frac{8\pi e^2}{\hbar} \frac{1}{\Omega} \sum_q \frac{1}{q^2} |< \mathbf{k}'|e^{i\mathbf{qr}}|\mathbf{k}>|^2 \mathrm{Im}\frac{1}{\varepsilon(q, E_{\mathbf{k}} - E_{\mathbf{k}'} + \mathrm{i}0^+)} \ .$$

(5.5)

The total excitation probability $W^{\mathrm{s}}(\mathbf{k}', \mathbf{k})$ of a crystal electron in the state $\mathbf{k}'$ by an electron $\mathbf{k}$ consists also of three terms,

$$W^{\mathrm{s}}(\mathbf{k}', \mathbf{k}) = W^{\mathrm{s}}_{\mathrm{e}}(\mathbf{k}', \mathbf{k}) + W^{\mathrm{s}}_{\mathrm{p}}(\mathbf{k}', \mathbf{k}) + W^{\mathrm{s}}_{\mathrm{c}}(\mathbf{k}', \mathbf{k}) \ . \tag{5.6}$$

Again, the contributions of single particle processes (e, c) can be obtained from (5.3, 1) with suitable transition matrix elements. $W^{\mathrm{s}}_{\mathrm{p}}(\mathbf{k}', \mathbf{k})$ is related to electronic transition processes with excitation and decay of plasmons. The physics underlying these processes will be discussed below.

In NFE metals the system of conduction electrons can be described to a first approximation within the free-electron gas picture. Neglecting exchange and correlation effects, the dielectric function is represented by the well-known expression in the random phase approximation (RPA), first proposed by *Lindhard* (1954),

$$\varepsilon^{\mathrm{L}}_1(x, y) = 1 + \frac{2\alpha r_{\mathrm{s}}}{\pi x^3}[1 + R(x, y) + R(x, -y)] \tag{5.7}$$

$$\varepsilon^{\mathrm{L}}_2(x, y) = \begin{cases} \frac{\alpha r_{\mathrm{s}}}{x^3} y & \text{for } y \le x(2 - x) \\ \frac{\alpha r_{\mathrm{s}}}{x^3}\left[1 - \left(\frac{x^2-y}{2x}\right)^2\right] & \text{for } x|2 - x| \le y \le x(2 + x) \\ 0 & \text{otherwise} \end{cases} \tag{5.8}$$

where

$$R(x, y) = \frac{1}{2x}\left[1 - \left(\frac{x^2 + y}{2x}\right)^2\right]\ln\left|\frac{x^2 + 2x + y}{x^2 - 2x + y}\right| \tag{5.9}$$

and $\alpha = (4/9\pi)^{1/3}$. Wave number and energy are measured in units of Fermi momentum and Fermi energy: $x = q/k_{\mathrm{F}}$ and $y = \hbar\omega/E_{\mathrm{F}}$, respectively. It is convenient to express the density n of the electron gas by the dimensionless parameter r_{s} which represents the radius of a sphere, in units of Bohr radius a_{B}, containing on the average one electron, $n = [4\pi(a_{\mathrm{B}}r_{\mathrm{s}})^3/3]^{-1}$. For the density corresponding to the conduction band of Al $r_{\mathrm{s}} = 2.07$.

In Fig. 5.2 the RPA excitation spectrum of the free-electron gas is shown. The region of individual excitations where $\varepsilon^{\mathrm{L}}_2 \ne 0$ is bounded by two parabolas which are obtained from the law of energy and momentum conservation. Besides the pair excitations for wave numbers below $x_{\mathrm{c}} = q_{\mathrm{c}}/k_{\mathrm{F}}$ (q_{c} is the cut-off wave number) there is a plasmon mode $y_{\mathrm{p}}(x) = \hbar\omega_{\mathrm{p}}(q)/E_{\mathrm{F}}$ which is determined by $\varepsilon^{\mathrm{L}}_1(x, y_{\mathrm{p}}(x)) = 0$. The finite value of this plasmon energy at zero wave number is related to the electron density by the well-known expression $\omega^2_{\mathrm{p}}(0) =$

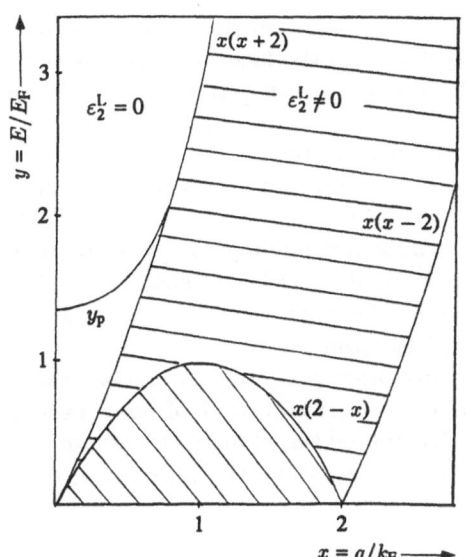

Fig. 5.2. Elementary excitation spectrum of the homogeneous free-electron gas (jellium) in the random phase approximation. For details see text

$4\pi n e^2/m$. For aluminum with $r_s = 2.07$ we obtain $\hbar\omega_p(0) = 15.7\,\text{eV}$, whereas the experimental value is $15.0\,\text{eV}$ (Raether 1980). This discrepancy can be attributed mainly to the contribution of interband processes as well as core polarization effects (Sturm 1982). In spite of this distinct influence of solid state effects on the plasmon properties, which takes place also in other NFE metals, for our purposes it seems to be sufficient to describe the direct excitation of conduction electrons by the impinging particle as well as the slowing-down of inner excited SE within the free-electron gas model in RPA. In this case the energy loss function can be decomposed into single particle processes and plasmon excitation processes as (Pines 1963)

$$\text{Im}\frac{1}{\varepsilon(q,\omega + i0^+)} = -\frac{\varepsilon_2(q,\omega)}{|\varepsilon(q,\omega)|^2} - \frac{\pi\hbar}{\left.\frac{\partial\varepsilon_1(q,\omega)}{\partial\omega}\right|_{\omega_p(q)}}\delta(\hbar\omega - \hbar\omega_p(q))\,\Theta(q_c - q)\,, \tag{5.10}$$

where $\varepsilon(q,\omega)$ is given by (5.7–9). Formula (5.10), if inserted into (5.5), leads directly to $W_e(\mathbf{k}',\mathbf{k})$ and $W_p(\mathbf{k}',\mathbf{k})$.

In the RPA there is no damping of bulk plasmons for $q < q_c$. However, in real metals a finite plasmon line width is observed. For aluminum the measured wave number dependent line width $\Gamma(q)$ (Krane 1978; Raether 1980) may be represented by a parabolic expression $\Gamma(q) = \Gamma_0 + \Gamma_2 \cdot (q/k_F)^2$ with $\Gamma_0 = 0.5\,\text{eV}$ and $\Gamma_2 = 3\,\text{eV}$, respectively. In the free-electron gas model the decay of plasmons for $q < q_c$ is possible only by higher-order processes, e. g., the creation of two electron-hole pairs or the simultaneous creation of one electron-hole pair and a plasmon with lower energy. However, these processes become more important only with increasing wave number. It is well established, especially in the small wave number region, that plasmon damping in simple metals is primarily determined by interband transitions (Paasch 1969; Sturm 1982).

In the case of particle-induced electron emission, plasmons can be excited by primary particles or by inner excited electrons. In the latter case the plasmon decay by interband processes is directly related to the excitation probability $W_p^s(\mathbf{k}', \mathbf{k})$.

In the description of the transport of inner excited SE we will neglect the contribution to transition and excitation probabilities determined by the interaction with core electrons. First, $W_c(\mathbf{k}', \mathbf{k})$ and $W_c^s(\mathbf{k}', \mathbf{k})$ differ from zero only if the energy difference $E_k - E_{k'}$ exceeds the lowest binding energy of the NFE metal (72.6 eV for the 2p-level in Al). Second, at the relevant energies (100 eV range) $W_c(\mathbf{k}', \mathbf{k})$ and $W_c^s(\mathbf{k}', \mathbf{k})$ are much smaller than the other contributions related to the interaction with conduction electrons.

Besides the different inelastic scattering processes discussed above, elastic scattering should be included in the description of transport of inner excited SE. Starting from a model of randomly distributed target atoms (randium) the transition probability $W^{(el)}(\mathbf{k}', \mathbf{k})$ can be expressed by the differential cross section $d\sigma/d\Omega$,

$$W^{(el)}(\mathbf{k}', \mathbf{k}) \sim N_{at} \frac{d\sigma(E_{k'}\vartheta)}{d\Omega} \delta(E_k - E_{k'}) , \qquad (5.11)$$

where N_{at} is the density of scatterers and ϑ the scattering angle $\vartheta = \angle(\mathbf{k}, \mathbf{k}')$.

Another basic quantity for the description of the transport of inner excited SE is the mean free path (mfp) of an electron in the state $\mathbf{k}$. In this case we must also take into account contributions from inelastic and elastic scattering. The mfp $l(E)$ is defined in the usual way by the transition probability from the state $\mathbf{k}$ to $\mathbf{k}'$. Using (5.4) and (5.11) we obtain

$$\frac{1}{l_{inel}(E)} = \frac{1}{v(E)} \sum_{k'} W(\mathbf{k}', \mathbf{k}) \qquad (5.12)$$

and

$$\frac{1}{l_{el}(E)} = \frac{1}{v(E)} \sum_{k'} W^{(el)}(\mathbf{k}', \mathbf{k}) , \qquad (5.13)$$

respectively. For the total mfp this means

$$\frac{1}{l(E)} = \frac{1}{l_{inel}(E)} + \frac{1}{l_{el}(E)} . \qquad (5.14)$$

5.2 General Form of the Transport Equation

The transport of SE from their point of excitation to the surface of the emitter can be described by the Boltzmann transport equation. Taking into account the stationary condition and the geometry of the problem (Fig. 4.1) this equation for the normalized density of excited electrons $N(x; E, \mathbf{\Omega})$ in the state $k(E, \mathbf{\Omega})$ at the depth x can be written as

$$- v(E)\cos\alpha\frac{\partial N(x;E,\Omega)}{\partial x} = S(x;E_0;E,\Omega) - \frac{v(E)}{l(E)}N(x;E,\Omega)$$

$$+ \int\int dE'\,d\Omega'\,W^\sigma(E,\Omega;E',\Omega')N(x;E',\Omega')\,. \tag{5.15}$$

The second term on the right-hand side denotes the number of particles thrown out of state $k(E,\Omega)$ by elastic and inelastic collisions. The mfp $l(E)$ is given by (5.14). The third term on the right-hand side of (5.15) denotes the number of electrons put into state $k(E,\Omega)$ by collisions. In calculating this term we have to take into account that both particles from $k'(E',\Omega')$ (with $E' > E$) and from the Fermi sphere can be put into the state $k(E,\Omega)$ (Fig. 5.1). The total transition function $W^\sigma(k,k') \sim W^\sigma(E,\Omega;E',\Omega')$ can be written as

$$W^\sigma(k,k') = W^{\sigma(\text{inel})}(k,k') + W^{(\text{el})}(k,k'). \tag{5.16}$$

According to (5.4, 6) the inelastic transition function is given by

$$W^{\sigma(\text{inel})}(k,k') = W(k,k') + W^{\text{s}}(k,k')\,. \tag{5.17}$$

The excitation function $S(x;E_0;E,\Omega)$ expresses the number of electrons in the state $k(E,\Omega)$ at the depth x created by the primary beam. Like the density $N(x;E,\Omega)$, this function is normalized to unit primary current. The spatial dependence stems in fact from the slowing-down of the primary particles on their path through the target.

In order to calculate the density of inner excited electrons at the surface $N(E,\Omega) = N(x = 0;E,\Omega)$, which is needed for the evaluation of the current density of outgoing electrons according to (4.7), we must solve (5.15) taking into account the boundary conditions for partial reflection (Puff 1964). Within such a comprehensive treatment of the transport problem in a semi-infinite geometry the last two stages of the three-step model mentioned in Chap. 3 cannot be separated.

The Boltzmann equation (5.15) can be applied to both emission phenomena (IIEE, SEE) considered here. Differences originate only in the excitation functions. Once more we will emphasize the role of the different important lengths (range of the incoming particle $R(E_0)$, escape depth L) introduced in Chap. 2. If the energy loss of the primary particle within the escape depth is negligible, which means $R(E_0) \gg L$, then the spatial dependence of the excitation function can be neglected. This is always fulfilled for light incident ions in the primary energy range considered here ($E_0 > 20\,\text{keV}$ for protons). However, for incident electrons the primary energies must be restricted to the range beyond the maximum in the yield curve (Fig. 2.2). To describe the SEE at smaller primary energies, e. g., in the region of the yield maximum, the slowing-down of the PE as well as the straggling of the primary beam must be taken into account. This can be done by solving the slowing-down transport equation for the PE using empirical range-energy relationships (Bennet and Roth 1972; Bindi, Lanteri, and Rostaing 1980a; Dubus, Devooght, and Dehaes 1987).

At very low primary energies, e. g., below $100\,\text{eV}$, true secondary and backscattered electrons cannot be distinguished, i. e., a separation of the primary and secondary electrons is impossible. Then stage 1 of the three-step model must be included in a complete solution of the electron transport problem with suitable boundary conditions. In this case only the current of PE at the surface is regarded as the source term.

It is beyond the scope of this work to give a complete microscopic theory of SEE for all primary energies. The main purpose is to present an almost complete microscopic description of different excitation as well as scattering processes (Rösler and Brauer 1988). This can be done only with considerable numerical effort. For that reason we will neglect the spatial dependence of the problem in further considerations. This leads to the above-mentioned restriction to the primary energy range ($E_0 \gtrsim 1 \, \text{keV}$) in the case of SEE.

5.3 Homogeneous Excitation

Neglecting the spatial dependence of the excitation function and changing from the density to

$$\psi(E, \boldsymbol{\Omega}) = \frac{v(E)}{l(E)} N(E, \boldsymbol{\Omega}) \,, \tag{5.18}$$

equation (5.15) may be written as

$$\psi(E, \boldsymbol{\Omega}) = S(E_0; E, \boldsymbol{\Omega}) + \int \int dE' \, d\boldsymbol{\Omega}' K^\sigma(E, \boldsymbol{\Omega}; E', \boldsymbol{\Omega}') \psi(E', \boldsymbol{\Omega}') \,. \tag{5.19}$$

The scattering function K^σ in (5.19) is simply related to the transition function W^σ in (5.15) by

$$K^\sigma(E, \boldsymbol{\Omega}; E', \boldsymbol{\Omega}') = \frac{l(E')}{v(E')} W^\sigma(E, \boldsymbol{\Omega}; E', \boldsymbol{\Omega}') \,. \tag{5.20}$$

In this model of slowing-down of excited electrons in an infinite medium the escape process must be inserted by the simple representation given in Chap. 4. With the solution of (5.19) the current density of outgoing electrons $j(E, \boldsymbol{\Omega})$ can be obtained from (4.7).

6. Scattering Functions and Mean Free Paths

The transport of inner excited SE is determined by the mean free paths and the scattering functions. In this chapter a detailed discussion of these basic quantities is given.

Previous investigations have shown that both inelastic scattering by the metal electrons and elastic scattering by the randomly distributed atoms of the target must be taken into account. The importance of elastic scattering in the case of SEE was first noted by *Kadyshevich* (1940). Both types of scattering processes determine the total mfp according to (5.14) as well as the total scattering function according to (5.16, 20) (written in suitable variables for further considerations),

$$K^{\sigma}(E, E', \cos\vartheta) = K^{\sigma(\text{inel})}(E, E', \cos\vartheta) + K^{(\text{el})}(E, E', \cos\vartheta) , \qquad (6.1)$$

where $\vartheta\angle(\boldsymbol{k}, \boldsymbol{k}')$ is the scattering angle.

Mean free paths and scattering functions are governed by the same fundamental interaction processes. This means that there are simple sum rules connecting the scattering functions and the corresponding mfp. In the past most of the explicit calculations of particle-induced electron emission used scattering functions and mfp based on different approximations for the underlying interaction processes. Recent calculations reveal the importance of a consistent description of transport of inner excited SE (Rösler and Brauer 1988).

6.1 Scattering Functions

6.1.1 Elastic Scattering. The description of elastic scattering starts from an atomic picture which considers the atoms of the solid as being randomly distributed. The transition probability from state $\boldsymbol{k}'$ to $\boldsymbol{k}$ is then, according to (5.11), obtained from the atomic one by multiplication with the density of atoms in the solid. With (5.11) the elastic scattering function can be written as

$$K^{(\text{el})}(E, E', \cos\vartheta) = N_{\text{at}} l(E) \frac{d\sigma(E, \vartheta)}{d\Omega} \delta(E - E') . \qquad (6.2)$$

Realistic scattering cross sections may be calculated by the partial wave method. In this case the differential scattering cross section is given by

$$\frac{d\sigma(E, \vartheta)}{d\Omega} = \frac{1}{k^2} \left| \sum_{l=0}^{\infty} (2l + 1) \sin\delta_l e^{i\delta_l} P_l(\cos\vartheta) \right|^2 , \qquad (6.3)$$

where δ_l is the phase shift suffered by the lth partial wave and P_l is the lth Legendre polynomial. With (6.3) we obtain for the scattering function

$$K^{(\text{el})}(E, E', \cos\vartheta) = N_{\text{at}} \frac{l(E)}{k^2} \delta(E - E') \sum_{l,l'=0}^{\infty} (2l + 1)(2l' + 1)$$
$$\times \sin\delta_l \sin\delta_{l'} \cos(\delta_l - \delta_{l'}) P_l(\cos\vartheta) P_{l'}(\cos\vartheta) . \qquad (6.4)$$

In general, the phase shifts δ_l are evaluated within a suitable muffin-tin approximation. For Al, numerical values of the phase shifts are obtained from computer analysis of LEED data based on a muffin-tin scheme using the $X\alpha$-approximation in describing exchange and correlation (Pendry 1980). There are only small differences between these values and the corresponding ones used in other calculations (Ganachaud and Cailler 1979) in the relevant energy range. The density of scatterers N_{at} is taken as $N_{\text{at}} = 6.07 \cdot 10^{22}$ cm^{-3}.

6.1.2 Inelastic Scattering. The inelastic scattering function $K^{\sigma(\text{inel})}(E, E', \cos\vartheta)$ is determined according to (5.17) by the transition probability from the state $\boldsymbol{k}'$

17

to $\boldsymbol{k}$ and the excitation probability of a crystal electron in the state $\boldsymbol{k}$ by an electron $\boldsymbol{k}'$. Both probabilities (5.4, 6) consist of three terms which are attributed to interaction processes with conduction electrons (e) and core electrons (c) and plasmon processes (p). Therefore, for the inelastic scattering function we may write

$$
\begin{aligned}
K^{\sigma(\text{inel})}(E, E', \cos\vartheta) &= \sum_i K_i^\sigma(E, E', \cos\vartheta) \\
&= \sum_i [K_i(E, E', \cos\vartheta) + K_i^s(E, E', \cos\vartheta)] ; \\
i &= \text{e}, \text{p}, \text{c} .
\end{aligned} \tag{6.5}
$$

According to the reasoning in Sect. 5.1 for the description of transport of inner excited SE we will neglect the contribution K_c^σ in (6.5).

The scattering function $K_e(E, E', \cos\vartheta)$ and $K_e^s(E, E', \cos\vartheta)$ can be obtained from the probabilities $W_e(\boldsymbol{k}.\boldsymbol{k}')$ and $W_e^s(\boldsymbol{k}, \boldsymbol{k}')$ defined in Sect. 5.1. Within the free-electron-gas model we get from (5.5, 10) and (5.1, 3) ($E = E_k$, $E' = E_{k'}$)

$$
K_e(E, E', \cos\vartheta) = \frac{e^2 m^2 k l(E')}{\pi^2 \hbar^4 k' |\boldsymbol{k}' - \boldsymbol{k}|^2} \frac{\varepsilon_2^{\text{L}}(|\boldsymbol{k}' - \boldsymbol{k}|, E' - E)}{|\varepsilon^{\text{L}}(|\boldsymbol{k}' - \boldsymbol{k}|, E' - E)|^2} \tag{6.6}
$$

and

$$
\begin{aligned}
K_e^s(E, E', \cos\vartheta) = &\frac{4 e^4 m^2 k l(E')}{\hbar^4 k'} \frac{1}{\Omega} \sum_{k''(<k_{\text{F}})} \frac{1}{|\boldsymbol{k} - \boldsymbol{k}''|^4} \\
&\times \frac{\Theta(E' + E'' - E - E_{\text{F}})}{|\varepsilon^{\text{L}}(|\boldsymbol{k} - \boldsymbol{k}''|, E - E'')|^2} \delta(E_{k''+k'-k} + E - E'' - E') ,
\end{aligned} \tag{6.7}
$$

respectively.

In order to reduce the numerical effort of explicit calculations it seems to be obvious that one should evaluate the scattering functions K_e and K_e^s using the Thomas-Fermi approximation for the dielectric function. This leads to analytical expressions. These Thomas-Fermi scattering functions were applied in previous works on SEE (Rösler and Brauer 1981a; 1981b) and IIEE (Rösler and Brauer 1984). However, a comparison of the electron-electron scattering functions calculated with different screening approximations (Rösler and Brauer 1988) shows that the Thomas-Fermi approximation greatly underestimates the electron-electron scattering rates for the transport of internal excited SE. Therefore, in further explicit calculations the frequency dependent RPA dielectric function will be used in (6.6, 7). It should be mentioned in this context that *Tung* and *Ritchie* (1977) in their calculation of electron slowing-down spectra in Al were the first to use the frequency dependent Lindhard dielectric function (5.7–9).

The second important contribution to the inelastic scattering function $K_p^\sigma(E, E', \cos\vartheta)$ is governed by plasmon processes. The first part of this scattering function $K_{\mathrm{p}}(E, E', \cos \vartheta)$ can be obtained from the plasmon contribution of the energy loss function (5.10). We have

$$K_{\mathrm{p}}(E, E', \cos \vartheta) = \frac{e^2 m^2 k l(E')}{2\pi\hbar^3 k' |k' - k|^2} \frac{\Theta(q_{\mathrm{c}} - |k' - k|)}{\omega_{\mathrm{p}}(|k' - k|) \frac{\partial \varepsilon_1^{\mathrm{L}}(|k' - k|, \omega)}{\partial \omega^2}\Big|_{\omega_{\mathrm{p}}(|k' - k|)}}$$
$$\times \delta(E' - E - \hbar\omega_{\mathrm{p}}(|k' - k|)) . \tag{6.8}$$

In order to determine the second part of the scattering function K_{p}^{σ} we must go beyond the free-electron-gas picture. $K_{\mathrm{p}}^{\mathrm{s}}(E, E', \cos \vartheta)$ is related to the excitation probability $W_{\mathrm{p}}^{\mathrm{s}}(k, k')$ introduced in Sect. 5.1. As discussed there the plasmon damping in NFE metals is primarily determined by interband processes. Therefore, these processes determine also the scattering function $K_{\mathrm{p}}^{\mathrm{s}}$. For the evaluation of this scattering function it seems to be convenient to use the extended zone scheme in describing interband transitions. Then the corresponding transition matrix element can be written as

$$< k' | \mathrm{e}^{-iqr} | k > = \sum_K \delta_{k, k'+q+K} B^K(k', k) , \tag{6.9}$$

where the Bloch integral $B^K(k', k)$ is defined by

$$B^K(k', k) = \frac{1}{\Omega} \int_\Omega d^3 r\, u_{k'}^*(r) u_k(r) \mathrm{e}^{iKr} . \tag{6.10}$$

$u_k(r)$ is the periodic part of the Bloch function and K denotes reciprocal lattice vectors. The other matrix element appearing in (5.1) is related to the electronic transition between states of higher energy. It should be calculated approximately using plane waves. With (6.9) we obtain from (5.1, 3, 20)

$$K_{\mathrm{p}}^{\mathrm{s}}(E, E', \cos \vartheta) = \frac{8e^4 m^2 k l(E')}{\hbar^4 k'} \frac{1}{\Omega} \sum_{K, q(<q_{\mathrm{c}})} \frac{|B^K(k', k' + q + K)|^2}{q^4 |\varepsilon(q, E' - E_{k'+q})|^2}$$
$$\times \Theta(E_{k'+q} - E_{\mathrm{F}}) \Theta(E_{\mathrm{F}} - [E + E_{k'+q} - E'])$$
$$\times \delta(E_{k'+q} + E - E' - E_{k'+q+K}) . \tag{6.11}$$

In NFE metals the electronic structure is well described in a model potential scheme. Wave functions and Bloch energies are given by perturbation theory with respect to this weak model potential V_{M}. The description of interband processes requires calculation of the electronic structure in the vicinity of zone boundaries with sufficient accuracy. This can be done by perturbation theory for nearly degenerate states (two-band model). The electronic states which appear in the transition matrix element can be written as

$$\psi_k(r) = \frac{1}{\sqrt{\Omega}}(a_0(k) + a_K(k)\mathrm{e}^{-iKr})\mathrm{e}^{ikr} . \tag{6.12}$$

The coefficients $a_0(k)$ and $a_K(k)$ are determined from the Schrödinger equation and the normalization condition. The periodic part of the Bloch function $u_k(r)$ is given by the bracket in (6.12). The Bloch energies $\hat{E}_k$ are given by the well-known square-root expression (Ashcroft and Mermin 1976)

$$\hat{E}_k = \frac{1}{2}\left[E_k + E_{k-K} \pm \sqrt{(E_k - E_{k-K})^2 + 4|V_K|^2} \right] . \tag{6.13}$$

Then, with (6.10) we obtain for the quantity $|B^K(k', k'' = k' + q + K)|^2$ which appears in the scattering function K_p^s

$$|B^K(k', k'')|^2 = |V_K|^2 \frac{(D_{k'} + D_{k''})^2}{(D_{k'}^2 + |V_K|^2)(D_{k''}^2 + |V_K|^2)} , \tag{6.14}$$

with $D_{k'} = E_{k'} - \hat{E}_{k'}$. V_K are the Fourier coefficients of the local model potential. Actual calculations were carried out with the model potential given by *Animalu* and *Heine* (1965).

In evaluating (6.11) the integration over the azimuthal angle in q-space is transformed to an integral over the energy loss $\Delta = E_{k'} - E_{k'+q}$. This Δ-integral can be performed using the resonance structure of $1/|\varepsilon(q,\omega)|^2$ in the frequency region about $\omega_p(q)$:

$$\frac{1}{|\varepsilon(q,\omega)|^2} = \frac{1}{\left(\left. \frac{\partial \varepsilon_1(q,\omega)}{\partial \omega} \right|_{\omega_p(q)} \right)^2} \frac{1}{(\omega - \omega_p(q))^2 + (\Gamma(q)/2\hbar)^2} . \tag{6.15}$$

In Fig. 6.1 we have plotted the plasmon part of the RPA excitation spectrum together with the plasmon damping $\Gamma(q)$ and the upper limit $\Delta_q = \hbar^2 q(2k' -$

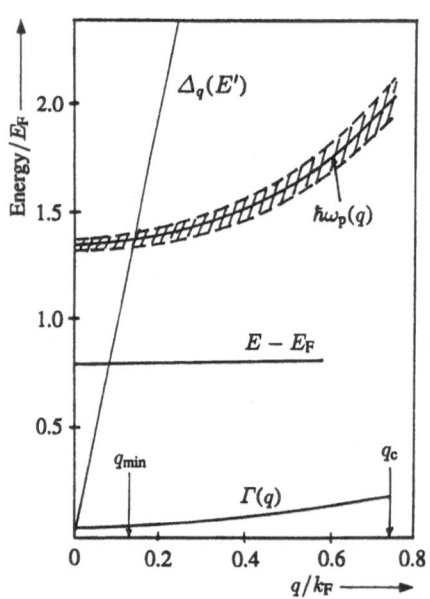

Fig. 6.1. Plasmon energy $\hbar\omega_p(q)$, plasmon damping $\Gamma(q)$, and limits of Δ-integration (Δ_q, $E - E_F$) as functions of wave number. q_c is the plasmon cut-off wave number, q_{min} is the minimum wave number related to decay of plasmons via interband processes in the case of plasmon excitation by an electron with energy E'

$q)/2m$ of Δ-integration. The relevant energy range for plasmon excitation is the region of width $\Gamma(q)$ about $\hbar\omega_{\mathrm{p}}(q)$ (shaded area in Fig. 6.1). The relative positions of different limits of Δ-integration with regard to this energy range determine the lower limit of q-integration $q_{\min}$. It should be noted that the q-integral is determined by the behavior at small q-values. Therefore, in order to reduce the numerical effort in evaluating (6.11) it is sufficient to use (6.14) for $k'' = k' + q + K$ in the limit of small q,

$$|B^K(k', k' + q + K)|^2 \approx |V_K|^2 \frac{(\hbar^4/m^2)(q, K)^2}{[(E_{k'+K} - E_{k'})^2 + 4|V_K|^2]^2} . \tag{6.16}$$

We also put $q = 0$ in the energy δ-function. Furthermore, supposing that $\hbar\omega_{\mathrm{p}}(q)$ is large compared with the energy gaps at the zone boundaries we can approximate the Bloch energies in the Θ- and δ-function in (6.11) by simple parabolic expressions.

Finally, in order to obtain a formula for polycrystalline materials, an average over all directions of K is performed. In explicit calculations for Al we took into account only the interband processes belonging to K_1 [111] and K_2 [200]. The corresponding Fourier coefficients of the model potential are $|V_{K_1}| = 0.24\,\mathrm{eV}$ and $|V_{K_2}| = 0.79\,\mathrm{eV}$, respectively (Animalu and Heine 1965). The plasmon damping $\Gamma(q)$ is given by the parabolic expression mentioned in Sect. 5.1.

6.2 Mean Free Paths

6.2.1 Elastic Scattering. According to (5.13) the elastic mfp is related to the transition probability $W^{(\mathrm{el})}(k', k)$. Using (5.11) in this formula we obtain with the total elastic scattering cross section

$$\sigma_{\mathrm{tot}}(E) = \int d\Omega \frac{d\sigma(E, \vartheta)}{d\Omega} \tag{6.17}$$

the simple relation

$$\frac{1}{l_{\mathrm{el}}(E)} = N_{\mathrm{at}} \sigma_{\mathrm{tot}}(E) . \tag{6.18}$$

For the total scattering cross section (6.17) we obtain from (6.3) the following representation in terms of phase shifts:

$$\sigma_{\mathrm{tot}}(E) = \frac{4\pi}{k^2} \sum_{l=0}^{\infty} (2l + 1) \sin^2 \delta_l . \tag{6.19}$$

The energy dependence of $l_{\mathrm{el}}(E)$ is shown in Fig. 6.2 for Al. It is important to note that the elastic mfp is of the order of a few Å in the relevant energy range.

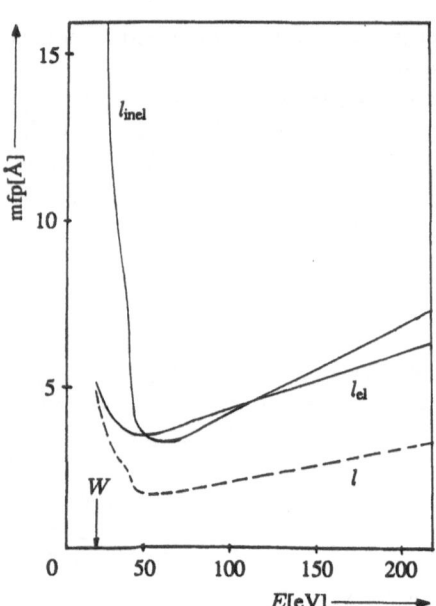

Fig. 6.2. Energy dependence of the total mean free path l (mfp) of electrons in Al (dashed curve). Comparison of elastic and inelastic mfp. The *arrow* indicates the vacuum level (calculated)

6.2.2 Inelastic Scattering. The inelastic mfp is defined by (5.12). According to (5.4), different scattering mechanisms contribute to l_{inel}

$$\frac{1}{l_{inel}(E)} = \frac{1}{l_e(E)} + \frac{1}{l_p(E)} + \frac{1}{l_c(E)} . \tag{6.20}$$

Using (5.5, 10) the contributions l_e and l_p related to interaction processes with the system of conduction electrons are obtained from (Quinn 1962)

$$\frac{1}{l_e(E)} = \frac{8\pi e^2 m}{\hbar^2 k} \frac{1}{\Omega} \sum_q \frac{1}{q^2} \frac{\varepsilon_2(q, E - E_{k+q})}{|\varepsilon(q, E - E_{k+q})|^2} \tag{6.21}$$

and

$$\frac{1}{l_p(E)} = \frac{4\pi^2 e^2 m}{\hbar k} \frac{1}{\Omega} \sum_{q(<q_c)} \frac{1}{q^2 \omega_p(q) \left.\frac{\partial \varepsilon_1(q,\omega)}{\partial \omega^2}\right|_{\omega_p(q)}}$$
$$\times \delta(E_{k+q} - E - \hbar\omega_p(q)) , \tag{6.22}$$

respectively.

There have been several attempts to calculate the inelastic mfp related to the free-electron gas beyond RPA (e.g. Ashley, Tung, and Ritchie 1979; Penn 1976) including damping effects or exchange and correlation. For our purposes it seems not necessary to calculate this contribution more accurately than obtained from (6.21, 22) using the RPA dielectric function (5.7–9). In this

22

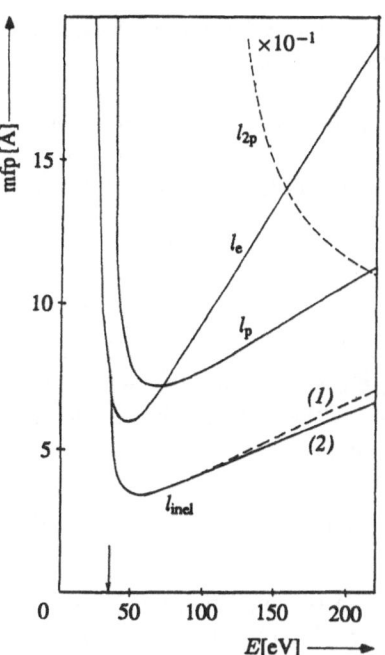

Fig. 6.3. Inelastic mfp of electrons vs. electron energy in Al. Comparison of different contributions: l_c is almost completely given by the $2p$-contribution l_{2p}. l_{inel} is the total inelastic mfp including (curve 2) and neglecting (curve 1) l_{2p}. The *arrow* indicates the plasmon threshold (calculated)

connection we will emphasize once more that for a consistent description of the transport process it is necessary to calculate the mfp and the scattering functions in the same approximations.

In Fig. 6.3 we have plotted the different contributions to l_{inel} together with the total inelastic mfp. In this figure l_c is also shown using the most reliable values given in the literature (Ashley, Tung, and Ritchie 1979). Because $l_c(E) \gg l_e, l_p$ in the 100 eV range we can neglect core excitations in the transport process, as mentioned before.

The comparison of elastic and inelastic contributions to the total mfp shown in Fig. 6.2 demonstrates that both contributions are of the same order of magnitude in the major part of the relevant energy range. However, especially at low energies l_{el} dominates the total mfp.

7. Excitation Functions

The interaction between the primary charged particles and the electron system of the metal leads to various possibilities of generating excited electrons. In this chapter the excitation functions concerning these different interaction processes are evaluated.

The parameters of the incident beam – mass, energy, and angle of incidence – enter the description of particle-induced electron emission through the excitation function. As mentioned in Chap. 5 the spatial dependence of the excitation rate will be neglected. Let us now regard the different excitation processes which are very similar for both SEE and IIEE. In the following we consider four mechanisms responsible for generating SE: excitation via screened particle-electron interaction (e), excitation by decay of plasmons generated by the incident particles (p), excitation of core electrons (c), and excitation by Auger processes (a) which immediately follow the excitation of inner-shell electrons. Therefore, the total excitation function can be written as

$$S(E_0; \boldsymbol{k}) = \sum_i S_i(E_0; \boldsymbol{k}) \,; \quad i = \mathrm{e}, \mathrm{p}, \mathrm{c}, \mathrm{a} \,. \tag{7.1}$$

A general formula for the excitation function may be obtained from the excitation probability (5.3) using (5.1). If $\boldsymbol{k}_0$, E_0, and v_0 denote the wave number, energy and velocity of the incoming particle, then the number of electrons thrown into the state $\boldsymbol{k}$ by the primary beam is given by

$$S(E_0; \boldsymbol{k}) = \frac{1}{v_0} W^{\mathrm{s}}_{\boldsymbol{k}_0, \boldsymbol{k}} = \frac{1}{v_0} \sum_{\boldsymbol{k}'_0, \boldsymbol{k}'(<k_{\mathrm{F}})} W_{\boldsymbol{k}_0 \boldsymbol{k}' \to \boldsymbol{k}'_0 \boldsymbol{k}} \,. \tag{7.2}$$

The expression for the transition probability can be simplified by the assumption that the impinging particle before ($\boldsymbol{k}_0$) and after ($\boldsymbol{k}'_0$) the scattering event is in a plane wave state. Then we have for the corresponding transition matrix element $< \boldsymbol{k}'_0 | \exp(\mathrm{i}\boldsymbol{q}\boldsymbol{r}) | \boldsymbol{k}_0 > = \delta_{\boldsymbol{k}'_0, \boldsymbol{k}_0 + \boldsymbol{q}}$. We arrive at the following formula:

$$S(E_0; \boldsymbol{k}) = \frac{64\pi^3 e^4}{\hbar v_0} \frac{1}{\Omega^2} \sum_{\boldsymbol{q}, \boldsymbol{k}'} \frac{1}{q^4} \frac{| < \boldsymbol{k} | \mathrm{e}^{-\mathrm{i}\boldsymbol{q}\boldsymbol{r}} | \boldsymbol{k}' > |^2}{|\varepsilon(q, E_0 - E_{\boldsymbol{k}_0 + \boldsymbol{q}})|^2}$$
$$\times \, \Theta(E_{\mathrm{F}} - E_{\boldsymbol{k}'}) \delta(E_0 + E_{\boldsymbol{k}'} - E_{\boldsymbol{k}_0 + \boldsymbol{q}} - E_{\boldsymbol{k}}) \,. \tag{7.3}$$

A factor of 2 results from the spin summation over the initial states in the Fermi sphere.

To proceed further, various approximations are necessary concerning the dielectric function, the transition matrix element, and the energies in order to obtain explicit expressions for the different excitation functions. Moreover, in order to simplify the considerations, perpendicular incidence of the primary beam will be assumed.

7.1 IIEE

In the case of the proton-induced electron emission, the energy and velocity of the incident ion are given by $E_0 = E^{\mathrm{i}}_{\boldsymbol{k}_0} = \hbar^2 k_0^2 / 2M_{\mathrm{p}}$ and $v_0 = \hbar k_0 / M_{\mathrm{p}}$. M_{p} denotes the proton mass. In the following we evaluate the excitation functions $S_i(E_0; \boldsymbol{k})$ with $i = \mathrm{e}, \mathrm{p}, \mathrm{c}$ starting from (7.3) and the excitation function $S_{\mathrm{a}}(E_0; \boldsymbol{k})$ using a simple picture of the underlying Auger process. After that, we compare these different contributions to the excitation rate. Finally, we discuss the problems connected with the influence of the system of target electrons on the charge state of the moving ion. In our simple description of excitation processes such effects will be neglected.

7.1.1 Excitation of Single Conduction Electrons. In order to calculate this contribution the conduction electrons will be described within the free-electron-gas model. Then the dielectric function is given by (5.7,8). With

$$< \mathbf{k}|(\exp(-i\mathbf{q}\mathbf{r})|\mathbf{k}' >= \delta_{\mathbf{k}',\mathbf{k}+\mathbf{q}}$$

we obtain from (7.3)

$$S_e(E_0;k) = \frac{64\pi^3 e^4 M_P}{\hbar k_0} \frac{1}{\Omega} \sum_{\mathbf{k}'} \frac{\Theta(E_F - E_{\mathbf{k}'})}{|\mathbf{k}' - \mathbf{k}|^4 |\varepsilon^L(|\mathbf{k}' - \mathbf{k}|, E_{\mathbf{k}'} - E_k)|^2}$$

$$\times \delta(E_{\mathbf{k}_0+\mathbf{k}'-\mathbf{k}}^i + E_k - E_{\mathbf{k}_0}^i - E_{\mathbf{k}'}) . \tag{7.4}$$

One angular integral can be carried out by utilizing the δ-function. This leads to restrictions on the excitation angle $\theta = \angle(\mathbf{k}_0, \mathbf{k})$

$$\cos\theta_1 \leq \cos\theta \leq \text{Min}[\cos\theta_2, 1] \tag{7.5}$$

with

$$\cos\theta_{1,2} = \frac{1}{2kk_0}\left[(k^2 - k_F^2)\left(\frac{M_P}{m} + 1\right) \mp 2k_F\sqrt{k_0^2 - \frac{M_P}{m}(k^2 - k_F^2)}\right] . \tag{7.6}$$

The condition for the occurrence of excitation is given by $\cos\theta_1 < 1$. We obtain an upper boundary E_m for the excitation from the conduction band,

$$E_m = \frac{\hbar^2}{2m}\left[\frac{2k_0 + k_F\left(\frac{M_P}{m} - 1\right)}{\frac{M_P}{m} + 1}\right]^2 . \tag{7.7}$$

In Fig. 7.1 the primary energy dependence of this quantity is shown for different light ions. With increasing ion mass there is a lowering of E_m for a given primary energy. The restriction to excitation energies smaller than E_m is important only for small primary energies[1].

With regard to the condition (7.5) the excitation function $S_e(E_0; E, \cos\theta) \sim S_e(E_0; \mathbf{k})$ must be calculated numerically. Starting from (7.4) we have

$$S_e(E_0; E, \cos\theta) = \frac{k}{\pi^3 e^2 a_B^3 k_0 |k_0 - k|}\left(\frac{M_P}{m}\right)^2$$

$$\times \Theta(\cos\theta_2 - \cos\theta)\Theta(\cos\theta - \cos\theta_1)$$

$$\times \int_{k_{min}}^{k_F} k' \, dk' \int_0^{2\pi} \frac{d\varphi}{|\mathbf{k}' - \mathbf{k}|^4 |\varepsilon^L(|\mathbf{k}' - \mathbf{k}|, E' - E)|^2} , \tag{7.8}$$

---

[1] The terms "small" and "high" primary energies in the following discussion pertain to $E_0 \lesssim 60$ and $E_0 \gtrsim 500$ keV, respectively.

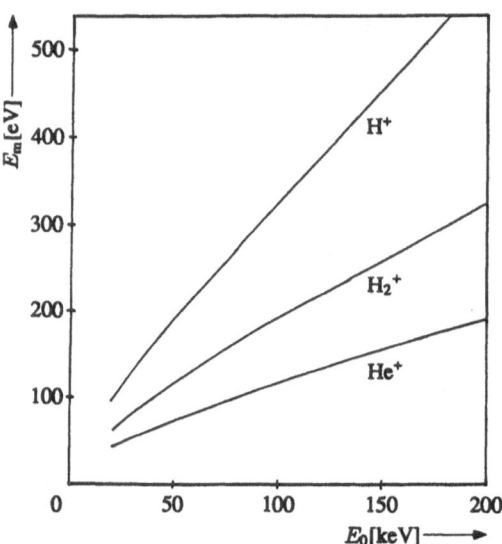

Fig. 7.1. Upper boundary for the excitation of single conduction electrons as a function of ion energy E_0 for different ions (calculated)

where

$$k_{min} = \frac{\left| |\boldsymbol{k}_0 - \boldsymbol{k}| - \sqrt{k_0^2 + k^2(M_p/m)^2 - 2k_0 k(M_p/m)\cos\theta} \right|}{(M_p/m) - 1} \tag{7.9}$$

and φ denotes the polar angle of $\boldsymbol{k}'$ with respect to $\boldsymbol{k}_0 - \boldsymbol{k}$.

In Fig. 7.2 the angular dependence of the excitation by screened proton-electron scattering is shown at low and high primary energies. At $E_0 = 40\,\text{keV}$ for increasing excitation energies the excitation takes place preferably in the direction of the primary beam. At high primary energies, however, the angular distribution is perpendicular to the direction of the primary beam at all relevant excitation energies.

The energy-dependent excitation function

$$4\pi S_{e,0}(E_0; E) = \int d\Omega\, S_e(E_0; E, \cos\theta) \tag{7.10}$$

is shown in Fig. 7.3 for low and high primary energies. In all cases there is a distinct peak around 35 eV above the bottom of the conduction band. This peak is connected with the resonant behavior of $1/|\varepsilon(q,\omega)|^2$ due to the penetration of the plasmon mode in the pair continuum of the RPA excitation spectrum (Fig. 5.2).

7.1.2 Excitation by Decay of Plasmons. In the following we consider the excitation of conduction electrons by decay of plasmons generated by the incident ion. As mentioned in Sect. 5.1 the plasmon damping in normal metals is dominated by interband processes. Higher-order processes within the free-electron-gas model responsible for plasmon damping will be neglected. The excitation

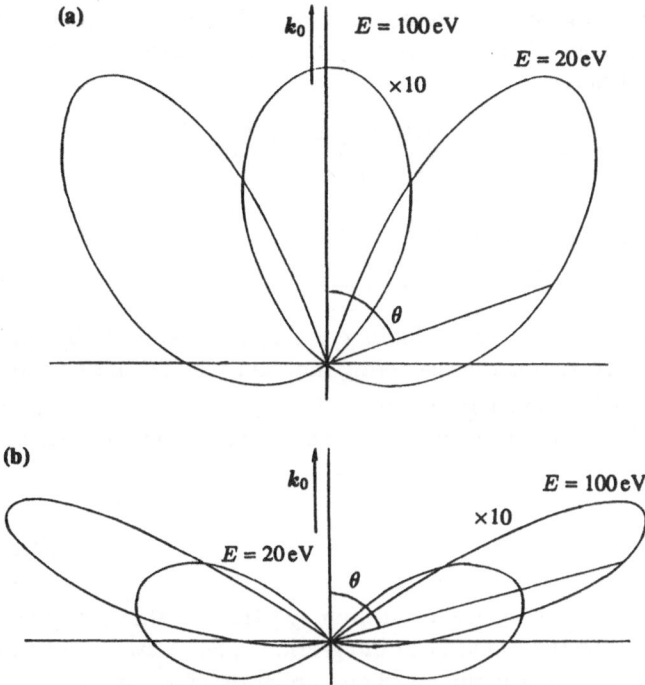

Fig. 7.2. Angular dependence of excitation by dynamically screened ion-electron scattering for $E = 20\,\mathrm{eV}$ and $100\,\mathrm{eV}$, respectively. (a) $E_0 = 40\,\mathrm{keV}$ and (b) $E_0 = 500\,\mathrm{keV}$. k_0 is the wave vector of the incident proton and θ is the excitation angle (calculated)

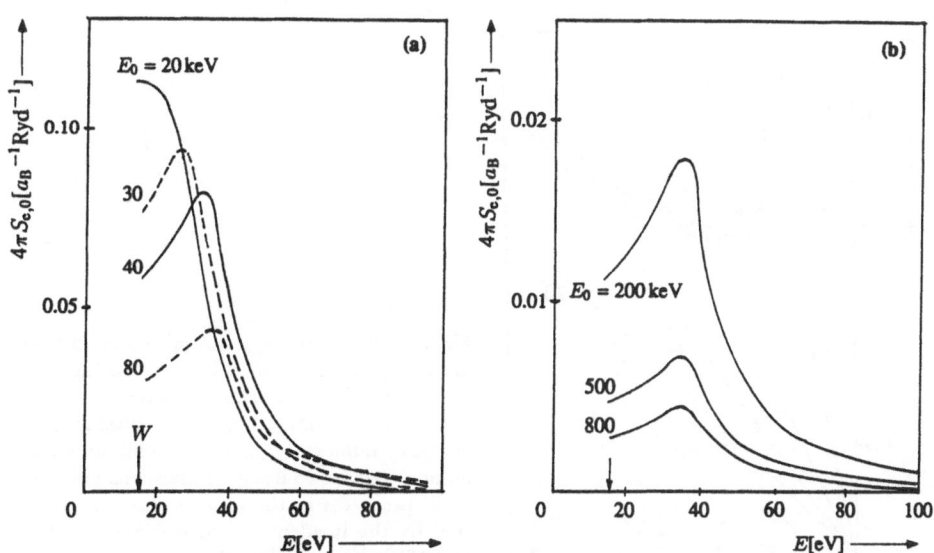

Fig. 7.3. Energy dependent excitation function for dynamically screened ion-electron scattering at (a) low and (b) high primary energies. The *arrow* indicates the vacuum level (calculated)

rate by interband processes can be calculated from (7.3). According to (6.9) the interband matrix element is represented by the Bloch integral. Then, by analogy to (6.11), we obtain

$$S_{\mathrm{p}}(E_0; \boldsymbol{k}) = \frac{64\pi^3 e^4 M_{\mathrm{p}}}{\hbar^2 k_0} \frac{1}{\Omega} \sum_{K, q(<q_c)} \frac{|B^K(\boldsymbol{k}, \boldsymbol{k}+\boldsymbol{q}+\boldsymbol{K})|^2}{q^4 |\varepsilon(q, E_{k_0}^{\mathrm{i}} - E_{k_0+q}^{\mathrm{i}})|^2}$$

$$\times \Theta(E_{\mathrm{F}} - [E_{k_0+q}^{\mathrm{i}} - E_{k_0}^{\mathrm{i}} + \hat{E}_k])$$

$$\times \delta(E_{k_0+q}^{\mathrm{i}} + \hat{E}_k - E_{k_0}^{\mathrm{i}} - \hat{E}_{k+q+K}) . \tag{7.11}$$

The mathematical procedure for evaluating $S_{\mathrm{p}}(E_0; \boldsymbol{k})$ is almost the same as that discussed in Sect. 6.1.2 for the scattering function $K_{\mathrm{s}}^{\mathrm{s}}$. The integrations over the momentum transfer and energy transfer $\Delta = E_{k_0}^{\mathrm{i}} - E_{k_0+q}^{\mathrm{i}}$ are restricted to the plasmon part of the excitation spectrum. The maximum energy transfer is given by

$$\Delta_q(E_0) = \frac{\hbar^2 q}{2M_{\mathrm{p}}}(2k_0 - q) . \tag{7.12}$$

In Fig. 7.4 the plasmon part of the excitation spectrum is shown together with plasmon damping and the limits of Δ-integration. The excitation of conduction electrons by plasmon decay via interband processes can only take place if the curve of maximum energy transfer intersects the region of the damped plasmon for $q < q_c$. In this way we obtain a minimum primary energy E_0^{min} for plasmon excitation. This energy is given by (M is the ion mass)

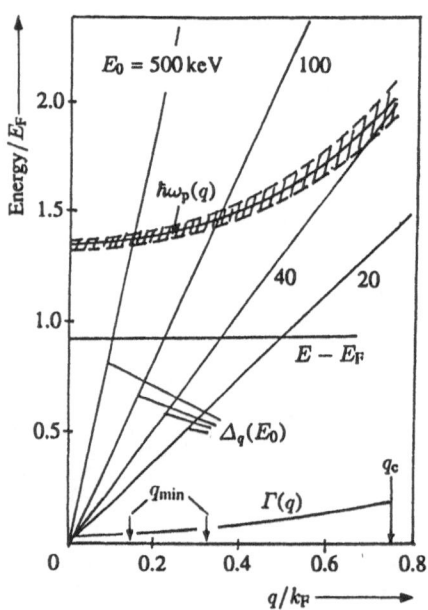

Fig. 7.4. Plasmon energy $\hbar\omega_{\mathrm{p}}(q)$, plasmon damping $\Gamma(q)$, and limits of Δ-integration ($\Delta_q(E_0)$, $E - E_{\mathrm{F}}$) for different primary energies as a function of wave number. q_{c} is the plasmon cutoff wave number. q_{min} is the minimum wave number related to decay of plasmons via interband processes in the case of plasmon excitation by the incident ion with energy E_0 (500 and 100 keV) (calculated)

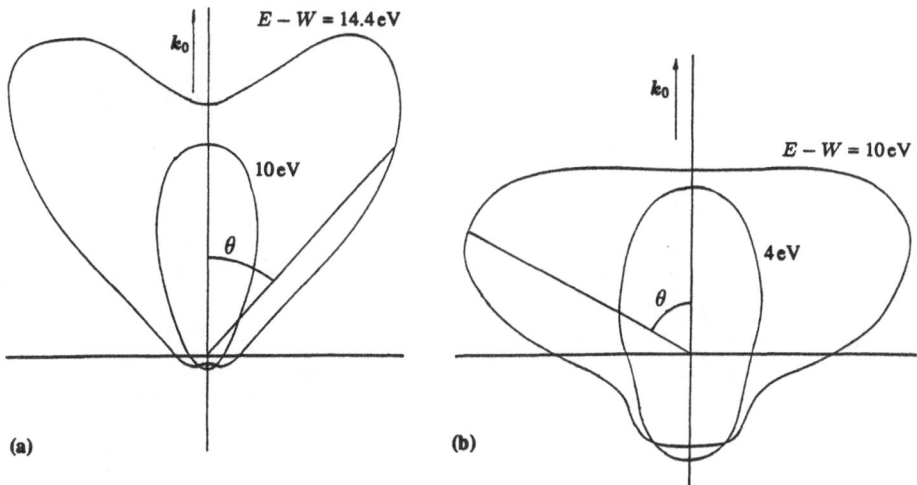

Fig. 7.5. Angular dependence of the excitation by plasmon decay for different secondary energies. (a) $E_0 = 60\,\text{keV}$ and (b) $E_0 = 500\,\text{keV}$. k_0 is the wave vector of the incident proton and θ is the excitation angle (calculated)

$$E_0^{\text{min}} = E_{\text{F}} \left(\frac{2 + q_{\text{c}}/k_{\text{F}}}{2} \right)^2 \frac{M}{m} \,. \tag{7.13}$$

For different projectile ions we obtain the following values: $E_0^{\text{min}} \approx 40\,\text{keV}$ for H^+, $E_0^{\text{min}} \approx 80\,\text{keV}$ for H_2^+, and $E_0^{\text{min}} \approx 1600\,\text{keV}$ for Ar^+.

The excitation via plasmon decay turns out to be anisotropic, especially at low primary energies as shown in Fig. 7.5. The excitation takes place preferably in the direction of the primary beam.

In Fig. 7.6 the energy dependent excitation function is shown for different primary energies of proton impact. There are remarkable qualitative differences for excitation at low and high primary energies. The formula for the excitation function (7.11) contains a factor $1/q^4$ resulting from the square of the Fourier transform of the Coulomb potential. Then, at high values of E_0 important contributions to the excitation rate stem from small momentum transfers $q \geq q_{\text{min}}$ (Fig. 7.4). Depending on the position of the lower limit of the Δ-integration, $E - E_{\text{F}}$, relative to the plasmon line the excitation function begins to drop at the energy $E_{\text{F}} + \hbar\omega_{\text{p}}$ ($q \approx 0$). At low values of E_0 the excitation rate is governed by momentum transfers $q_{\text{c}} > q > q_{\text{min}}$ where q_{min} approaches q_{c} with E_0 lowering to E_0^{min}. Then, the possible excitation energies are determined by the dispersion of the plasmon energy. The excitation function begins to decrease at the energy $E_{\text{F}} + \hbar\omega_{\text{p}}(q_{\text{min}})$.

In the evaluation of the excitation function by decay of plasmons for Al we have taken into account only the interband processes belonging to the reciprocal lattice vectors $K_1[111]$ and $K_2[200]$. However, in NFE metals there is a finite contribution to plasmon damping also from the interband process related to the

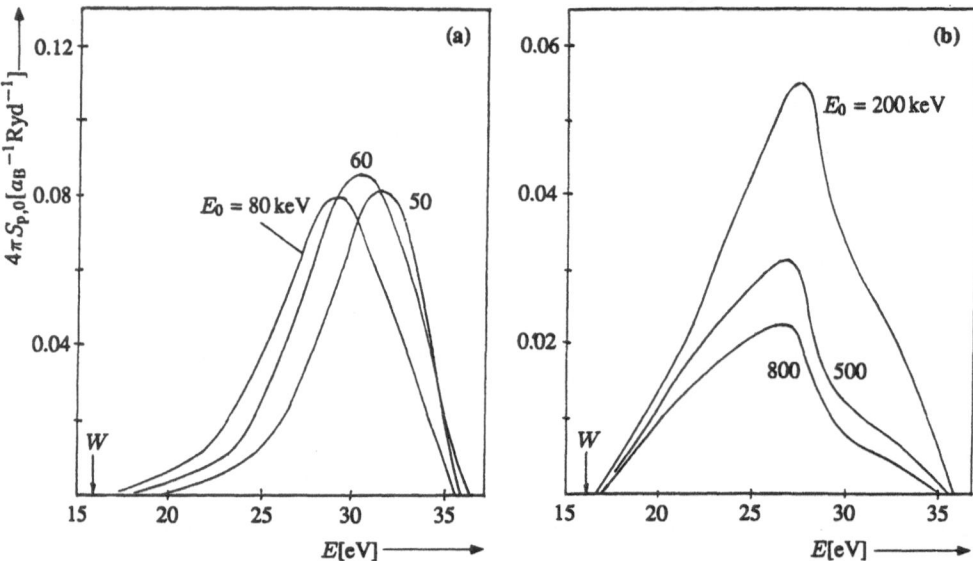

Fig. 7.6. Energy dependent excitation function by plasmon decay at (a) low and (b) high primary energies. The *arrow* indicates the vacuum level (calculated)

reciprocal lattice vector $K_3[220]$ (Sturm 1976, 1977, 1982; Rösler and Brauer 1981b). Therefore, we expect also the excitation of conduction electrons by this interband process. The possible excitation energies for this type of interband process are restricted to a very small interval ($< 0.5\,\mathrm{eV}$). This leads to an unrealistically narrow peak in the energy distribution of emerging electrons (Rösler and Brauer 1981b), in contradiction to experiment. At present there is no unambiguous explanation of this contradiction. A possible reason for the suppression of the K_3-contribution in (7.11) is related to the band structure. If we take into account the realistic band structure of Al (Levinson, Greuter, and Plummer 1983) in the relevant energy range, then the unrealistic K_3-peak which is a consequence of the description of the band structure within the simple model of nearly free electrons would probably disappear.

7.1.3 Excitation of Core Electrons. In this case we neglect the screening in (7.3) by reason of the large frequency argument in the dielectric function. In the transition matrix element the core states are described by Bloch sums (Fischbeck 1966; Arendt 1969),

$$\psi_{k\nu}(\boldsymbol{r}) = G^{-3/2} \sum_{\boldsymbol{R}} \mathrm{e}^{\mathrm{i}\boldsymbol{k}\boldsymbol{R}} \varPhi_{\nu}(\boldsymbol{r} - \boldsymbol{R}) \tag{7.14}$$

and the excited states by orthogonalized plane waves,

$$\psi_{k}(\boldsymbol{r}) = \frac{1}{\sqrt{\Omega}} \mathrm{e}^{\mathrm{i}\boldsymbol{k}\boldsymbol{r}} - \sum_{\boldsymbol{k}'\nu',\boldsymbol{K}} b_{\boldsymbol{k}\nu'} \psi_{\boldsymbol{k}'\nu'}(\boldsymbol{r}) \delta_{\boldsymbol{k},\boldsymbol{k}'+\boldsymbol{K}} \ , \tag{7.15}$$

where k is non-reduced, ν is the band index, and G^3 is defined by $G^3 = \Omega/\Omega_0$ where Ω_0 is the volume of the unit cell. $\Phi_\nu(r - R)$ is the atomic wave function centered at the lattice point R. For the core states (7.14) the Bloch energies are approximately given by the corresponding atomic levels: $E_{k\nu} \approx E_\nu = E_{nl}$ (n, l are the atomic quantum numbers). For the energies of the excited states (7.15) we use simple parabolic expressions. The coefficients $b_{k\nu}$ in (7.15) are determined by the demand for orthogonality of this state to all states $\psi_{k\nu}$ (7.14). This means

$$b_{k\nu} = \frac{1}{\sqrt{\Omega_0}} \int_{\Omega_0} e^{ikr} \Phi_\nu^*(r) \, d^3r \; . \tag{7.16}$$

The excitation function can be written in appropriately chosen variables

$$S_c(E_0; E, \cos\theta) = \frac{k}{\pi^3 e^2 a_B^3 k_0^2} \left(\frac{M_p}{m}\right)^2 \sum_\nu \int_{k_0-k_0'}^{k_0+k_0'} \frac{dq}{q^3} \int_0^{2\pi} d\varphi |B_\nu(k,q)|^2 \; , \tag{7.17}$$

where

$$B_\nu(k,q) = \int_{\Omega_0} e^{iqr} \Phi_\nu(r) \left[\frac{e^{-ikr}}{\sqrt{\Omega_0}} - \sum_{\nu'} b_{k\nu'}^* \Phi_{\nu'}^*(r) \right] d^3r \; , \tag{7.18}$$

and φ is the polar angle of k_0' with respect to k_0. $q = k_0 - k_0'$, where $k_0' = k_0\sqrt{1 - (E - E_\nu)/E_0}$.

In evaluating (7.17) we will only take into account the excitation of L-shell electrons. Due to the large binding energy of the $1s$-electron the excitation from the K-shell will be neglected. The binding energies of the L-shell of Al (measured up to the Fermi level) are 117.6 eV ($2s$) and 72.6 eV ($2p$), respectively. Actual calculations are carried out using the Herman-Skillman functions (Herman and Skillman 1963) for the radial part of the atomic wave function.

Fig. 7.7 shows the angle dependent excitation function for different secondary energies at low and high primary energies. At low secondary energy ($E \lesssim 50$ eV) the excitation is nearly isotropic, but with increasing energy it occurs preferably in the forward direction. In Fig. 7.8 the energy dependent excitation function is shown for different primary energies. The excitation rate is almost completely determined by that from $2p$ core states. In order to explain this we have plotted in Fig. 7.8 for $E_0 = 60$ keV the contribution of $2s$ core states to the excitation rate.

7.1.4 Auger Excitation. At the impact of energetic ions on the target by excitation of core electrons (discussed in the preceding section) inner shell vacancies are produced which within a very short time ($\lesssim 10^{-12}$ s) are filled with electrons from outer shells or from the conduction band. Here we are not interested in the special features of different transition processes which make the Auger elec-

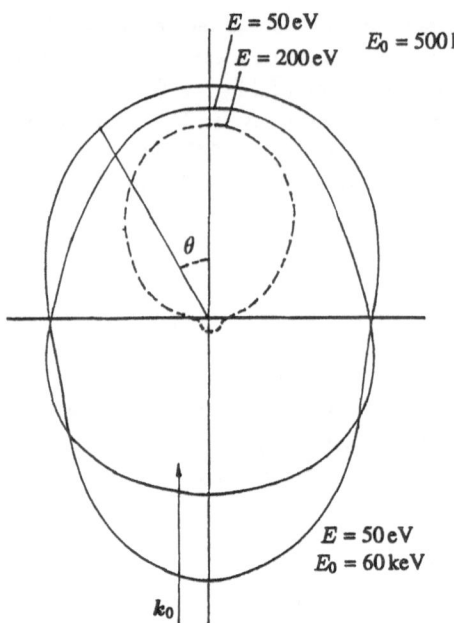

$E = 50\,\text{eV}$
$E = 200\,\text{eV}$
$E_0 = 500\,\text{keV}$

θ

$E = 50\,\text{eV}$
$E_0 = 60\,\text{keV}$

k_0

Fig. 7.7. Angular dependence of core excitation at different secondary energies. $E_0 = 60\,\text{keV}$ and $E_0 = 500\,\text{keV}$. k_0 is the wave vector of the incident proton and θ is the excitation angle (calculated)

tron spectroscopy a powerful method for investigating the electronic structure of solids. We consider the influence of the production of more or less monoenergetic electrons with isotropic distribution at relatively high excitation energies on the electron emission phenomenon. We are particularly interested in their influence on the distribution of emerging electrons at low energies as well as on the total electron yield.

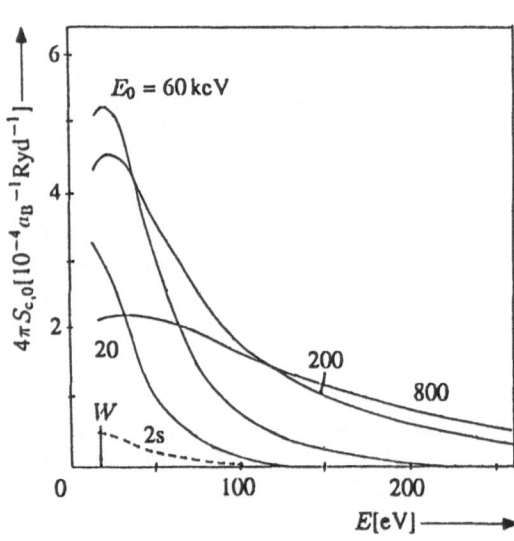

$4\pi S_{c,0}[10^{-4} a_B^{-1} \text{Ryd}^{-1}]$

$E_0 = 60\,\text{kcV}$

20

200

800

W

2s

0   100   200

$E[\text{eV}]$

Fig. 7.8. Energy dependent excitation function of core excitations from 2s and 2p core states at different primary energies. For $E_0 = 60\,\text{keV}$ the contribution of the 2s core states to the excitation rate is shown (dashed curve). The *arrow* indicates the vacuum level (calculated)

As in the case of direct excitation from the core levels of Al, where L-shell excitation is considered, we take into account only the filling of L-shell vacancies by conduction electrons. Auger processes (Coster-Kronig transitions) in which an initial vacancy is filled by an electron from the same shell are neglected.

Because both states of the Auger process are located in the conduction band the Auger line width is roughly twice the Fermi energy. However, due to the complicated structure of the density of states the actual Auger line width is considerably smaller. It will be considered as an input parameter in actual calculations.

We employ a simple model for calculating the Auger excitation function. We make the ansatz

$$S_{\mathrm{a}}^{\nu}(E_0; E) = A_{\nu} \frac{1}{\pi} \frac{\Gamma_{\nu}}{(E - E_{\nu}^{\mathrm{a}})^2 + \Gamma_{\nu}^2/4} , \tag{7.19}$$

where Γ_{ν} and E_{ν}^{a} are determined by width and position of the experimentally observed Auger peak. For the 2p-level in Al we have $\Gamma_{2p} \approx 10\,\mathrm{eV}$ and $E_{2p}^{\mathrm{a}} = W + 67\,\mathrm{eV}$. The coefficient A_{ν} is determined by the basic assumption that the number of excited electrons via Auger mechanism is equal to the number of excited core electrons or inner shell vacancies, respectively. Therefore,

$$\int S_{\mathrm{c},0}^{\nu}(E_0; E)\,dE = \int S_{\mathrm{a}}^{\nu}(E_0; E)\,dE . \tag{7.20}$$

The integral on the left-hand side is simply related to the contribution of the core states $\nu(2p, 2s)$ to the reciprocal mean free path

$$\frac{1}{l_{\nu}(E_0)} = 4\pi \int S_{\mathrm{c},0}^{\nu}(E_0; E)\,dE . \tag{7.21}$$

Then, from the condition (7.20) we obtain with (7.19)

$$A_{\nu} = \frac{1}{4\pi l_{\nu}(E_0)} . \tag{7.22}$$

A more simplified version of the excitation function can be obtained replacing the distribution in (7.19) by a δ-function,

$$S_{\mathrm{a}}^{\nu}(E_0; E) = A_{\nu} \delta(E - E_{\nu}^{\mathrm{a}}) . \tag{7.23}$$

For the 2p-core states both approximations of the Auger excitation function lead practically to the same results for yield and low energy distribution of the escaping electrons.

7.1.5 Comparison of Different Excitation Mechanisms. Obviously, all of the discussed excitation processes occur simultaneously. Therefore, it seems useful to compare the various excitation rates. In Fig. 7.9 the different energy dependent excitation functions are represented for $E_0 = 500\,\mathrm{keV}$.

Excitation by decay of plasmons is essentially restricted to energies E below $\hbar\omega_{\mathrm{p}}(q_{\mathrm{c}}) + W$ ($\approx 35\,\mathrm{eV}$ for Al). The number of electrons excited from the

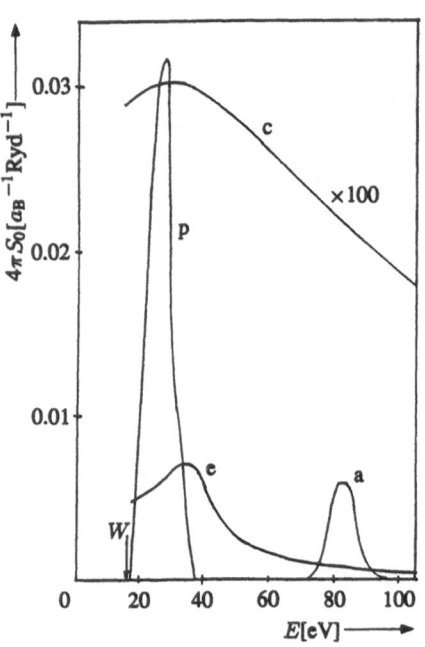

Fig. 7.9. Energy dependence of different excitation functions at $E_0 = 500\,\text{keV}$. Excitation – by decay of plasmons (p); – of single conduction electrons (e); – of core electrons (c); Auger excitation (a). The *arrow* indicates the vacuum level (calculated)

conduction band by direct scattering processes is smaller than those excited by plasmon decay at low E. However, in this case there is a large number of electrons with higher energies. As we will see later, these electrons are more effective for the formation of the electron cascade, and, therefore, for the number of emerging electrons. The number of electrons excited from inner shells is negligible at low energies. However, at higher E (above 200 eV) the number of electrons excited in this way exceeds the excitation rate from the conduction band. Because of the enhanced effectivity of energetic electrons we expect that also the excitation from core states leads to an important contribution to the number of escaping electrons. The energy dependent excitation function due to Auger processes is given in our simple theoretical model by (7.19) or (7.23). In this case the number of directly emitted electrons is small compared with the contributions from the other excitation mechanisms. Nevertheless, this quasi-monoenergetic excitation influences the emission of electrons by means of the electron cascade. The strength of the excitation rate is determined according to (7.22) by the core contribution l_c to the mfp of the impinging particle.

7.1.6 Critical Discussion of the Simple Interaction Model. Our theoretical considerations start from the simple picture of interacting point charges. In this case we can use the golden rule expression (7.3) in order to calculate the excitation functions. A general theory applicable to ions other than H^+ is beyond the scope of the present work. But also in the case of a positive point charge moving in the electron system of the target there are complications which are related to the possibility of formation of bound electron states.

For heavier ions we must take into account the effect of projectile electrons. The yield obtained in this case is larger than expected from a simple scaling law

$$\gamma_{\text{ion}}(E_0) = \gamma_{\text{H}+}(E_0 \cdot M_{\text{p}}/M) \,, \qquad (7.24)$$

which follows from the theoretical treatment of the ion as a single charged point charge of mass M. Starting from the excitation function (7.8, 11, 17) one can prove that all these expressions depend on $k_0/M \sim v_0$ only. This means that different single charged ions moving with equal velocity produce the same electron yield. With respect to the primary energy as the reference variable (7.24) follows immediately. Only for D^+ ($M = 2M_{\text{p}}$) the scaling law (7.24) is exactly fulfilled as has been confirmed by experiments (Baragiola, Alonso, and Oliva Florio 1979; Baragiola et al. 1979). This would be expected because there is no projectile electron in this case.

Our treatment of the excitation of electrons is based on the assumption that on their path through the target the charge state of the ion remains unchanged. However, in reality, the velocity dependent charge state is determined by the formation of bound states. The equilibrium charge state for H^+ and He^+ moving in a homogeneous electron gas was calculated by *Guinea, Flores,* and *Echenique* (1982). These charge states are obtained in terms of the processes of capture and loss for the level bound to the ion. A more elaborate investigation including dynamical screening and atomic-like transitions of electrons is given by *Echenique, Flores,* and *Ritchie* (1988). At high enough ion velocity $v_0 \geq 2$ [a. u.] (E_0 above $100 \,\text{keV}$ for H^+), neither screening by conduction electrons nor the detailed structure of the crystal affect the charge states in metals. In this case the ratio of the probability of finding a bare proton to that for a H^0 atom is much larger than 1. Then, the excitation rate can be obtained with sufficient accuracy from (7.3). At low ion velocity ($v_0 < 1.5$ [a. u.] or E_0 below $60 \,\text{keV}$ for H^+) the mentioned ratio decreases drastically to values smaller than 1. This can be interpreted as a reduction of the effective charge of the proton. Thus, in the region of low primary energies the employment of (7.3) leads to an overestimation of the number of excited electrons.

7.2 SEE

As in the foregoing, (7.3) is the starting point for the evaluation of excitation functions. Energy and velocity of the PE are given by $E_0 = \hbar^2 k_0^2/2m$ and $v_0 = \hbar k_0/m$, respectively.

7.2.1 Excitation of Single Conduction Electrons. In this case the different steps performed in Sect. 7.1.1 leading to (7.8) must be repeated. The excitation rate can be obtained utilizing

$$S_e(E_0; E, \cos\theta) = \frac{k}{\pi^3 e^2 a_B^3 k_0 |k_0 - k|} \Theta(\cos\theta_2 - \cos\theta)\Theta(\cos\theta - \cos\theta_1)$$

$$\times \int_{k_{\min}}^{k_F} k'\, dk' \int_0^{2\pi} \frac{d\varphi}{|k' - k|^4 |\varepsilon^L(|k' - k|, E' - E)|^2}, \tag{7.25}$$

where

$$k_{\min} = \frac{k(k_0 \cos\theta - k)}{|k_0 - k|} \tag{7.26}$$

and

$$\cos\theta_{1,2} = \frac{k^2 - k_F^2 \mp k_F \sqrt{k_0^2 - k^2 + k_F^2}}{k_0 k}. \tag{7.27}$$

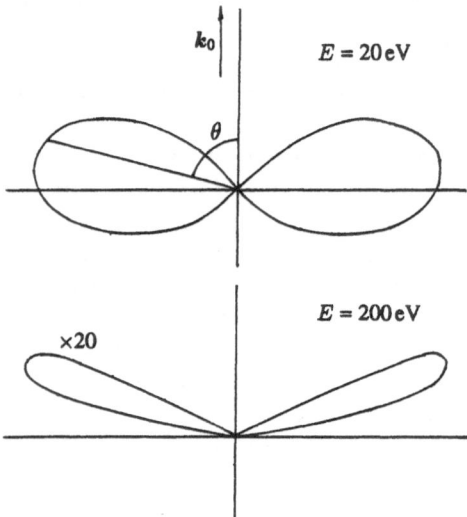

Fig. 7.10. Angular dependence of the excitation by dynamically screened electron-electron scattering for $E = 20\,\text{eV}$ and $E = 200\,\text{eV}$. $E_0 = 2\,\text{keV}$. k_0 is the wave vector of the primary electron and θ is the excitation angle (calculated)

In Fig. 7.10 the angular dependence of the excitation by screened electron-electron scattering is shown at $E_0 = 2\,\text{keV}$ for different secondary energies. In fact the excitation takes place nearly perpendicular to the direction of the primary beam. With increasing excitation energy a small preference of the forward direction can be observed. There are some similarities with the proton-induced excitation at high primary energies.

The energy dependent excitation function $4\pi S_{e,0}(E_0; E)$ is shown in Fig. 7.11 at different primary energies. Again we find qualitatively the same behavior as in the case of proton-induced electron emission (Fig. 7.3b).

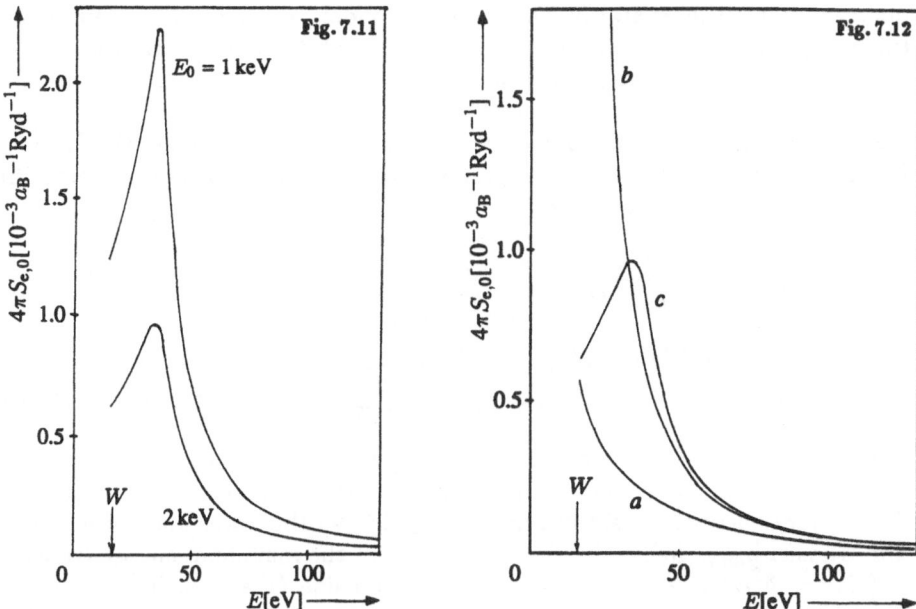

Fig. 7.11. Energy dependent excitation function by dynamically screened electron-electron scattering at different primary energies. The *arrow* indicates the vacuum level (calculated)

Fig. 7.12. Energy dependent excitation function by screened electron-electron scattering: Thomas-Fermi (*a*), unscreened, Streitwolf function (*b*), and RPA (*c*). $E_0 = 2\,\mathrm{keV}$ (calculated)

It is interesting to compare the energy dependent excitation rate calculated from (7.25) using different screening approximations. Neglecting screening in (7.25) we obtain the excitation function first derived by *Streitwolf* (1959)

$$S_e(E_0; E, \cos\theta) = \frac{k}{\pi^2 e^2 a_B^3 k_0} \Theta(\cos\theta_2 - \cos\theta)\Theta(\cos\theta - \cos\theta_1)$$
$$\times \frac{k_F^2(k_0^2 + k^2 - 2k_0 k \cos\theta) - (k_0 k \cos\theta - k^2)^2}{(k^2 - k_F^2)^2(k_0^2 + k^2 - 2k_0 k \cos\theta)^{3/2}}.$$

(7.28)

Using the Thomas-Fermi approximation for the dielectric function $\varepsilon(q, \omega) \approx \varepsilon(q, 0) \approx 1 + k_{TF}^2/q^2$, where k_{TF} ($k_{TF}^2 = 4k_F/\pi a_B$) is the Thomas-Fermi momentum, the excitation function was evaluated by *Chung* and *Everhart* (1977). Also in this case all integrations in (7.25) can be done analytically. In Fig. 7.12 we compare these approximations with the full dynamical RPA calculation which must be performed numerically. Obviously, the Streitwolf function overestimates the excitation rate at low secondary energies. At high excitation energies screening is ineffective and the excitation rates calculated in RPA and neglecting screening ($\varepsilon = 1$) approach each other. The Thomas-Fermi screening

37

underestimates the excitation rate at all secondary energies. Therefore, from this comparison we can observe the importance of dynamical screening in the evaluation of the excitation function S_e.

7.2.2 Excitation by Decay of Plasmons. In this case we can also use the formula for the excitation rate related to proton impact with minor amendments. In close analogy to (7.11) we have

$$S_p(E_0; k) = \frac{64\pi^3 e^4 m}{\hbar^2 k_0} \frac{1}{\Omega} \sum_{K, q(<q_c)} \frac{1}{q^4} \frac{|B^K(k, k+q+K)|^2}{|\varepsilon(q, E_{k_0} - E_{k_0+q})|^2}$$
$$\times \Theta(E_F - [E_{k_0+q} - E_{k_0} + \hat{E}_k])$$
$$\times \delta(E_{k_0+q} + \hat{E}_k - E_{k_0} - \hat{E}_{k_0+q+K}) \, . \qquad (7.29)$$

The further steps of the evaluation of (7.29) bear a great resemblance to the case of proton impact at high primary energies. It follows, for instance, that at the primary energies considered here ($E_0 \geq 1\,\text{keV}$) the lower limit of momentum transfer $q_{\min}$, defined by the intersection of the maximum energy transfer $\Delta_q(E_0) = \hbar^2 q(2k_0 - q)/2m$ with the plasmon line (Fig. 7.4), is very small. Therefore, we expect vast similarities in the behavior of energy and angular distribution of excitation rates for SEE and proton-induced electron emission at high primary energies.

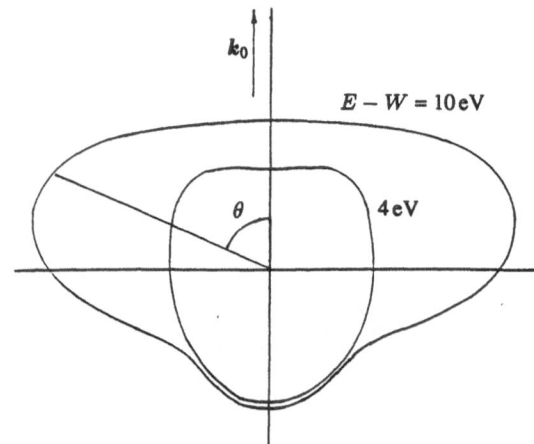

k_0

$E - W = 10\,\text{eV}$

θ $4\,\text{eV}$

Fig. 7.13. Angular dependence of the excitation by decay of plasmons for different secondary energies. $E_0 = 2\,\text{keV}$. k_0 is the wave vector of the primary electron and θ is the excitation angle (calculated)

In Figs. 7.13 and 7.14 we have plotted angular dependence and energy distribution of the excitation rate by plasmon decay, respectively.

7.2.3 Excitation of Core Electrons and Excitation by Auger Processes. In order to obtain the excitation function we can use (7.17) without the mass dependent prefactor

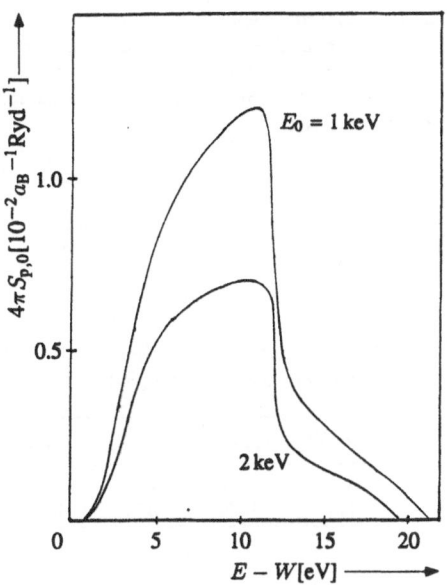

Fig. 7.14. Energy dependent excitation function by plasmon decay at different primary energies (calculated)

$E_0 = 1\,\text{keV}$

$2\,\text{keV}$

$$S_c(E_0; E, \cos\theta) = \frac{k}{\pi^3 e^2 a_B^3 k_0^2} \sum_\nu \int\limits_{k_0 - k_0'}^{k_0 + k_0'} \frac{dq}{q^3} \int\limits_0^{2\pi} d\varphi\, |B_\nu(\boldsymbol{k}, \boldsymbol{q})|^2 \qquad (7.30)$$

where $k_0' = \sqrt{k_0^2 - k^2 + (2m/\hbar^2)E_\nu}$.

The angle dependent excitation function shown in Fig. 7.15 reveals the same behavior as that discussed in Sect. 7.1.3 for proton impact at high primary energies. The same is true for the energy dependent excitation function shown in

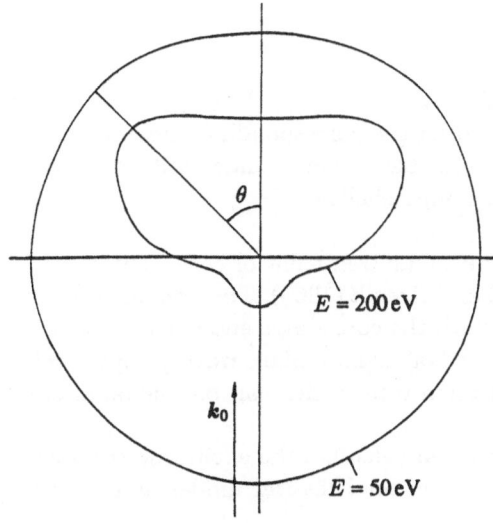

θ

$E = 200\,\text{eV}$

$\boldsymbol{k}_0$

$E = 50\,\text{eV}$

Fig. 7.15. Angular dependence of core excitation at different secondary energies. $E_0 = 2\,\text{keV}$. $\boldsymbol{k}_0$ is the wave vector of the primary electron and θ is the excitation angle (calculated)

39

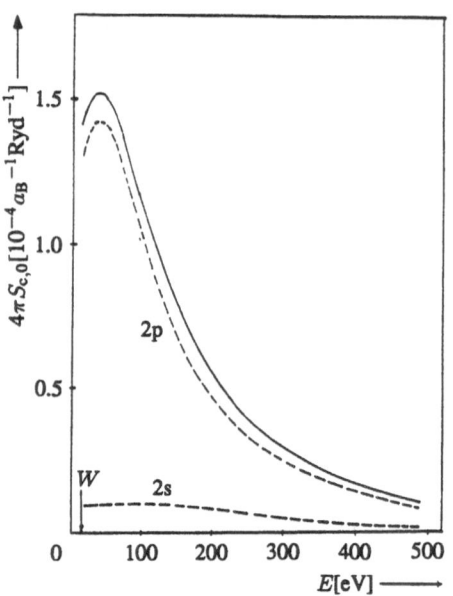

Fig. 7.16. Energy dependent excitation function of core excitation. Contributions from $2p$ and $2s$ core states are shown separately. $E_0 = 1\,\text{keV}$. The *arrow* indicates the vacuum level (calculated)

Table 7.1. Core contributions to the mfp at different primary energies

E_0 [keV]	l_{2s} [Å]	l_{2p} [Å]	l_c [Å]
1	1810	270	235
2	3170	390	320

Fig. 7.16, from which we can state that the excitation rate is almost completely determined by that from the $2p$ core states.

The excitation of electrons by Auger processes is closely related to the excitation of core electrons. In Sect. 7.1.4 we have proposed a simple model for Auger excitation which is also applicable for electron impact. According to (7.22) the strength of this excitation is governed by the corresponding core contributions to the mfp of the PE. These contributions to the mfp are calculated by means of (7.21) using the core excitation function (7.30). In determining l_c only the L-shell contributions are taken into account:

$$\frac{1}{l_c} = \frac{1}{l_{2p}} + \frac{1}{l_{2s}} \tag{7.31}$$

Numerical values are given in Table 7.1

These values are somewhat larger than the corresponding ones calculated by *Ashley, Tung,* and *Ritchie* (1979) on the basis of an atomic model developed by *Manson* (1972) for the ionization of inner shells.

7.2.4 Atomic Model of Core Excitations. Our treatment of core excitations is based on the work of *Fischbeck* (1966) and *Arendt* (1969). The crystal electrons are described by a Bloch scheme, in which the core states and excited states of electrons are given by Bloch sums and orthogonalized plane waves, respectively. This leads to expressions for the excitation rates which can be evaluated only with considerable numerical effort.

Besides this elaborate treatment we can calculate the excitation from core states within an atomic picture. Using the semiclassical model of *Gryzinski*

(1965) the differential cross section for an energy transfer ΔE from an electron with energy E_0 to an electron in a core state n, l with energy $E_{n,l}$ can be written as

$$\frac{d\sigma_{nl}(\Delta E, E_0, E_{nl})}{d\Delta E} = \frac{\sigma_0}{(\Delta E)^3} g(x, y) , \tag{7.32}$$

where $\sigma_0 = \pi e^4 = 651.4$ (eV Å)^2 and

$$g(x, y) = \frac{1}{x} \left(\frac{x}{1+x}\right)^{3/2} \left(1 - \frac{y}{x}\right)^{1/(1+y)}$$
$$\times \left[y \left(1 - \frac{1}{x}\right) + \frac{4}{3} \ln(2.718 + \sqrt{x - y})\right] \tag{7.33}$$

with $x = E_0/E_{nl}$ and $y = \Delta E/E_{nl}$.

The excitation function, which is isotropic by construction, can be obtained from the differential cross section by multiplying by the occupation number n_{nl} of the level n, l and the number of atoms per unit volume N_{at} $(E = E_{nl} + \Delta E)$

$$4\pi S_0^{nl}(E_0; E) = N_{\text{at}} n_{nl} \frac{d\sigma_{nl}}{d\Delta E} . \tag{7.34}$$

In Fig. 7.17 we compare the energy dependent excitation function (7.34) with our more elaborate calculation.

7.2.5 Comparison of Different Excitation Mechanisms. Figure 7.18 summarizes the different energy dependent excitation functions for $E_0 = 2\,\text{keV}$. We obtain

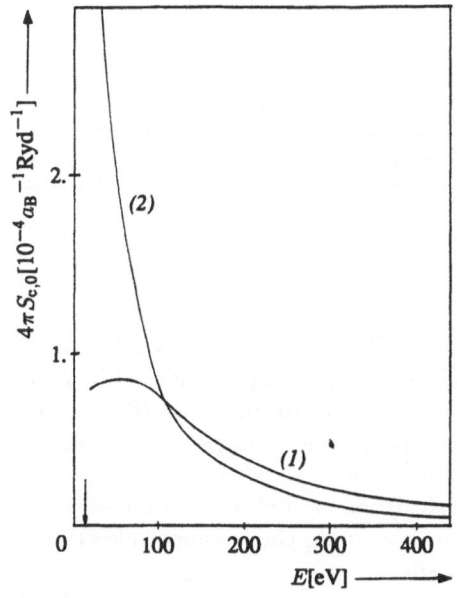

Fig. 7.17. Energy dependent excitation function of core electrons at $E_0 = 2\,\text{keV}$. Comparison of our calculation (*1*) with the model of *Gryzinski* (1965) (*2*). The *arrow* indicates the vacuum level

41

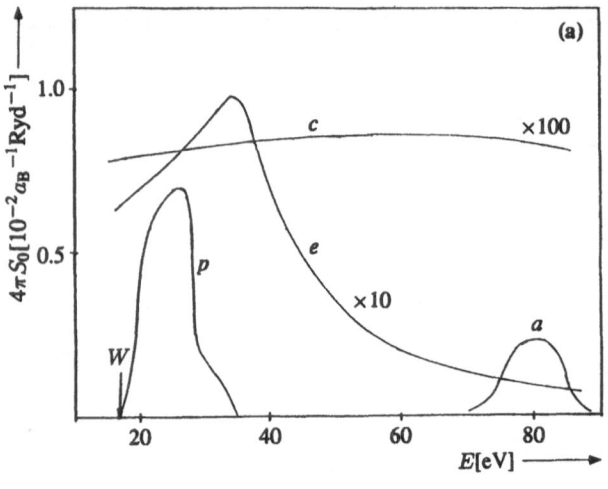

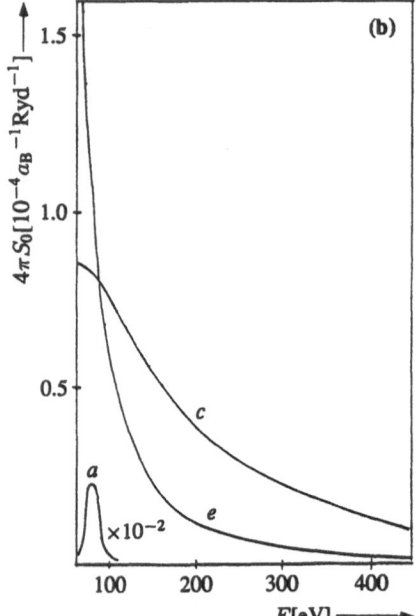

Fig. 7.18. Energy dependence of different excitation functions (see figure caption Fig. 7.9) at low (a) and high (b) secondary energies. $E_0 = 2\,keV$. The *arrow* indicates the vacuum level (calculated)

the same qualitative behavior of different excitation mechanisms relative to each other as in the case of proton impact at high primary energies (Fig. 7.9). From such a picture it is hardly possible to make a conclusion about the relative importance of various excitation mechanisms for the electron yield. It should be noted from Fig. 7.18b, however, that the excitation of core electrons yields the dominant contribution to the number of electrons at high excitation energies (above 100 eV). This would be significant in relation to the enhanced effectivity of energetic electrons via the electron cascade.

8. Solution of the Transport Equation

A general discussion of the different steps for solving the Boltzmann equation (5.19) is given in this chapter. The behavior of various excitation and scattering functions discussed in the preceding chapters require careful consideration.

8.1 General Discussion

The angular dependence of the problem will be treated by expansion in terms of Legendre polynomials. Such a procedure is advantageous if only a small number of expansion terms is necessary for describing the excitation function with sufficient accuracy. Then we may write

$$\psi(E, \mathbf{\Omega}) = \sum_{l=0}^{\infty} (-1)^l \psi_l(E) P_l(\cos \theta) \,, \tag{8.1}$$

$$S(E_0; E, \cos \theta) = \sum_{l=0}^{\infty} S_l(E_0; E) P_l(\cos \theta) \,, \tag{8.2}$$

$$K^\sigma(E, E', \cos \vartheta) = \sum_{l=0}^{\infty} K_l^\sigma(E, E') P_l(\cos \vartheta) \,. \tag{8.3}$$

Inserting these expansions into (5.19) we obtain immediately a set of independent Volterra integral equations of the second kind:

$$\psi_l(E) = (-1)^l S_l(E_0; E) + \int_E dE' K_l^\sigma(E, E') \psi_l(E') \,. \tag{8.4}$$

To proceed further it is useful to separate from the coefficient

$$K_l^\sigma(E, E') = 2\pi \int_{-1}^{+1} d\cos \vartheta K^\sigma(E, E', \cos \vartheta) P_l(\cos \vartheta) \tag{8.5}$$

the contribution of elastic scattering processes. Using (6.4) we may write

$$K_l^{(\text{el})}(E, E') = \hat{K}_l(E) \delta(E - E') \tag{8.6}$$

with

$$\hat{K}_l(E) = 2\pi N_{\text{at}} \frac{l(E)}{k^2} \sum_{l',l''=0}^{\infty} (2l' + 1)(2l'' + 1) \sin \delta_{l'} \sin \delta_{l''}$$
$$\times \cos(\delta_{l'} - \delta_{l''}) T_{l,l',l''} \tag{8.7}$$

where

43

$$T_{l,l',l''} = \int\limits_{-1}^{+1} dx \cdot P_l(x) P_{l'}(x) P_{l''}(x) \, . \qquad (8.8)$$

With (8.6) equation (8.4) can be rewritten as

$$\psi_l(E) = \tilde{S}_l(E_0; E) + \int\limits_{E} dE' \tilde{K}_l^\sigma(E, E') \psi_l(E') \, . \qquad (8.9)$$

$\tilde{S}_l$ is related to the excitation function by

$$\tilde{S}_l(E_0; E) = \frac{(-1)^l S_l(E_0; E)}{1 - \hat{K}_l(E)} \qquad (8.10)$$

and $\tilde{K}_l^\sigma$ is determined by the scattering function for all inelastic processes

$$\tilde{K}_l^\sigma(E, E') = \frac{K_l^{\sigma(\mathrm{inel})}(E, E')}{1 - \hat{K}_l(E)} \, . \qquad (8.11)$$

In general, the energy integration in (8.9) must be extended up to the primary energy E_0. However, either the excitation rate decreases with increasing energy in such a way that a restriction to energies below a suitable maximum value $E_{\max}$ is sufficient (S_e, S_c) or the excitation rate itself is restricted to a finite energy interval (S_p, S_a). In every case the upper limit $E_{\max}$ is clearly below the primary energy E_0. In our explicit calculations for the excitation of single conduction electrons or from core levels we use a value slightly above 400 eV as an upper limit for $E_{\max}$.

The Boltzmann equation can be solved by standard numerical methods. Equation (8.9) is transformed in the usual way to a system of linear algebraic equations:

$$\psi_l(E_i) = \tilde{S}_l(E_0; E_i) + \sum_{j(>i)}^{i_{\max}} V_l(i, j) \psi_l(E_j) \, ; \quad i = 1, 2, \ldots, i_{\max} \, . \qquad (8.12)$$

Depending on the behavior of the excitation function, a suitable energy mesh size is chosen,

$$E_i = E_1 + (i - 1)\Delta E \, ; \quad i = 1, \ldots, i_{\max} \, ;$$
$$(E_1 = W, E_{i_{\max}} = E_{\max}) \, . \qquad (8.13)$$

The coefficients $V_l(i, j)$ in (8.12) are determined by the integration rule used for converting the integrals into sums as well as according to (8.11) by the scattering functions (elastic, inelastic). We have used the simple trapezoidal rule to convert the integrals to sums.

8.2 Monoenergetic Isotropic Excitation

We will discuss the simple case of a monoenergetic isotropic excitation ($l = 0$):

$$\tilde{S}_0(E_0; E) = A\delta(E - E_0) \, . \qquad (8.14)$$

This is of fundamental interest to understand the origin of the energy distribution of inner excited electrons at low energies. Such an excitation will be realized approximately by Auger processes, as mentioned in Sect. 7.1.4 [see (7.23)].

The treatment of the monoenergetic isotropic excitation is closely connected with the problem of solving the Boltzmann equation by a Green's function method. The general solution of (8.9) can be written in terms of Green's functions $G_l(E, E')$ as (Stolz 1959)

$$\psi_l(E) = \int\limits_{E}^{E_{max}} dE'\, G_l(E, E') \tilde{S}_l(E_0; E')\,.$$
(8.15)

These Green's functions must be determined by the solution of

$$G_l(E, E') = \delta(E - E') + \int\limits_{E}^{E_{max}} dE''\, \tilde{K}_l^\sigma(E, E'') G_l(E'', E')\,.$$
(8.16)

Note that the $G_l(E, E')$ depend only on the properties of the metal.

Here we are interested only in the component $G_0(E, E')$ which contains the most important information on the effect of scattering processes on the internal spectral distribution $\psi_0(E) = A G_0(E, E_0)$. As shown in Chap. 6 the scattering functions are given by complicated expressions containing the dielectric function. Therefore, in spite of the simple form of the inhomogeneity in (8.16) the solution of this equation is possible only by numerical methods.

For the δ-like excitation (8.14) we have plotted the energy distribution of the density of internal electrons $N(E) = [l(E)/v(E)]\psi_0(E)$ together with the energy distribution of emerging electrons in Fig. 8.1. As can be seen from this figure the density of internal electrons grows considerably with decreasing energy. The enhancement of the density at low energies is determined, in fact, by the electron-electron scattering. The process of slowing down energetic electrons can be explained by using the iterative solution procedure of (8.16). Every step in this procedure ($n = 0, 1, 2, \ldots$) corresponds to the inclusion of contributions from n-fold scattered excited electrons. In Fig. 8.2 we illustrate this process of accumulation of electron density distribution at low energies.

We can state that the restriction to unscattered and single-scattered electrons in the calculation of the emission characteristics as proposed by *Chung* and *Everhart* (1977) is not sufficient to describe the actual distribution of electron density. This should be not only the case for the δ-like excitation considered here but also for realistic excitation functions.

The energy distribution of emerging electrons $j(E)$ is shown in Fig. 8.1. Using (4.7) and (2.1) $j(E)$ is governed by the density and the escape factor according to

$$j(E) = \pi \left(1 - \frac{W}{E'}\right) v(E') N(E')\,.$$
(8.17)

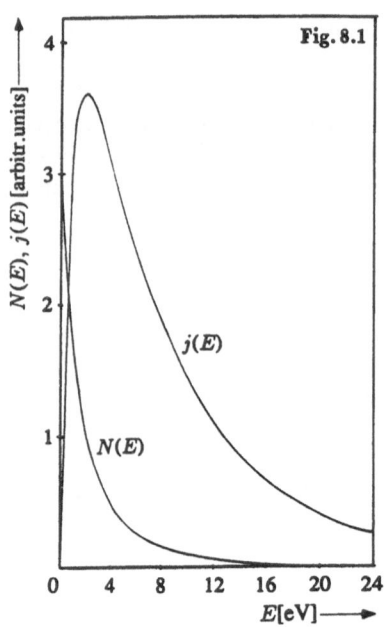
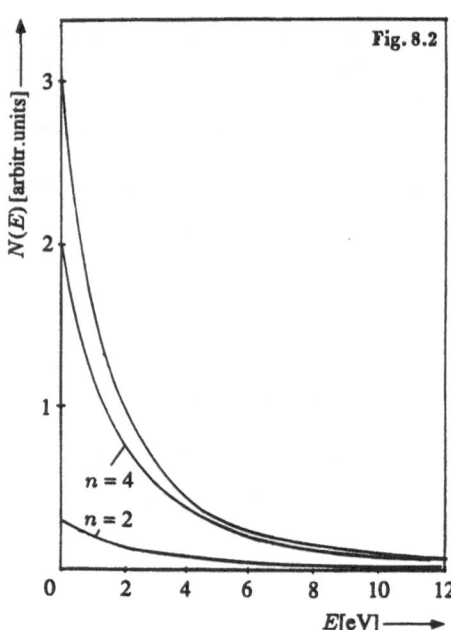

Fig. 8.1. Energy dependence of the density of inner SE, $N(E)$, and energy distribution of ejected electrons, $j(E)$ in the case of monoenergetic isotropic excitation. $E_0 = W + 67\,\text{eV}$ $= 82.9\,\text{eV}$ (calculated)

Fig. 8.2. Accumulation of the density distribution of inner SE at low energies. The number n at the different curves denotes the density distribution including n-fold scattered electrons. $E_0 = 82.9\,\text{eV}$ (calculated)

The general shape of the distribution is determined by the scattering properties of the system of target electrons and the escape process. This means that the position of the cascade maximum at $\approx 2\,\text{eV}$ as well as the half-width of the distribution are independent of the parameters of the primary beam. Special features of $j(E)$ which can be seen in the case of Al (Chap. 9) are attributed to peculiarities of the excitation processes.

9. Results for Aluminum and Comparison to Experimental Data

In this chapter the results of our calculations for different measurable quantities are presented and compared, as possible, with existing experimental data. The results for Al were obtained without fitting of parameters.

With the solution of (8.9) the different measurable quantities can be obtained from the general formula for the energy and angular dependent outer current density, by inserting into (4.7) the expansion (8.1) of $\psi(E, \boldsymbol{\Omega})$ in Legendre polynomials. Explicit formulas will be given only for the energy distribution of emerging electrons

$$j(E) = 2\pi l(E')\Theta(E' - W) \sum_{l=0}^{\infty} A_l(E')\psi_l(E') \qquad (9.1)$$

and the electron yield γ (or δ_p)

$$\gamma(\text{or } \delta_p) = 2\pi \sum_{l=0}^{\infty} \int_{W}^{W+50} dE' \, l(E')A_l(E')\psi_l(E') \qquad (9.2)$$

where

$$A_l(E') = \int_{\sqrt{\frac{W}{E'}}}^{1} dx \, x P_l(x) . \qquad (9.3)$$

For the transformation between outer and inner variables (4.1) is used.

9.1 IIEE

9.1.1 Energy and Angular Distribution. Up to now, zero energy in the distribution $j(E)$ denotes the vacuum level. According to (7.1) the total energy distribution of escaping electrons can be obtained by summing-up the contributions of the different excitation mechanisms. In Fig. 9.1 this distribution is shown for different primary energies.

The energy range in Fig. 9.1 is restricted to low energies (below 20 eV). At higher energies there are some peaks which are related to Auger processes (at $E_{2p}^a - W \approx 67\,\text{eV}$ for $2p$ and $E_{2s}^a - W \approx 112\,\text{eV}$ for $2s$ processes). These peaks which are clearly seen in the experimental spectra correspond to directly emitted electrons for this excitation mechanism. However, the contribution of these electrons to the total number of escaping electrons is negligible.

In accordance with the experiments at intermediate and high primary energies we found a maximum of the distribution $j(E)$ at $\approx 2\,\text{eV}$, the half-width of the distribution is $\approx 8.5\,\text{eV}$ and the plasmon shoulder at 11 to 12 eV. This energy position determined by the slope of the excitation function by plasmon decay is approximately given by $\hbar\omega_p(0) - \Phi$.

Until now there have been no measurements of the energy distribution at low primary energies. In this case our calculations predict a somewhat smaller half-width of the distribution as well as a shift of the plasmon shoulder to higher energies. The energy of the latter determined by plasmon dispersion is approximately given by $\hbar\omega_p(q_{\min}) - \Phi$. The position of the so-called cascade maximum at low energies ($\approx 2\,\text{eV}$) is a common feature of the energy distributions at all primary energies.

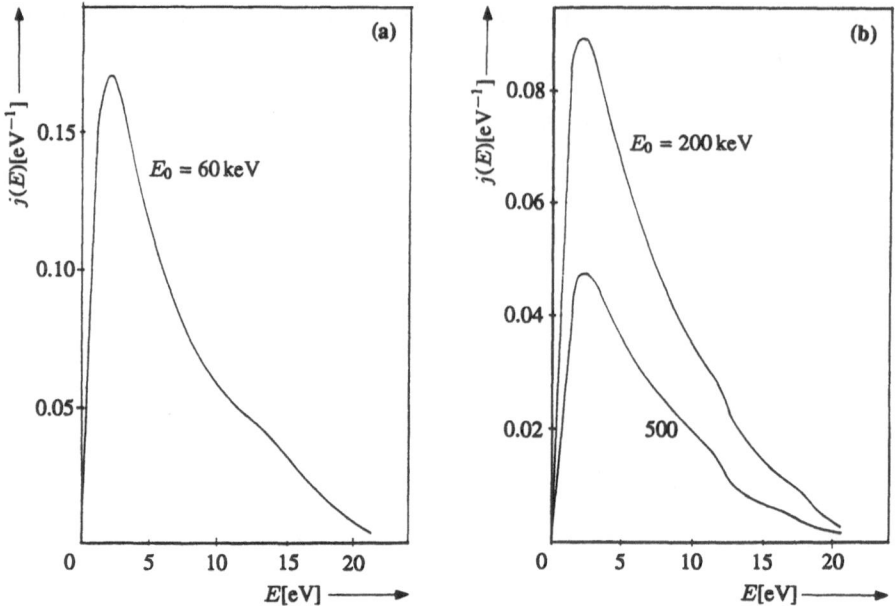

Fig. 9.1. Energy distribution of ejected electrons at (a) low ($E_0 = 60\,\mathrm{keV}$) and (b) intermediate and high primary energies (calculated)

The angular distribution of escaping electrons obtained by our calculations follows almost a cosine law. In the case of IIEE this angular distribution has not yet been measured.

9.1.2 Yield. As shown in Fig. 2.3 experimental values for the electron yield exist over a large range of primary energies. Therefore, we have extended existing calculations (Rösler and Brauer 1988) up to proton energies of about 10 MeV.

In Fig. 9.2 we have plotted the primary energy dependence of the different contributions to the electron yield γ. At low primary energies the contributions from inner shell excitations as well as from Auger excitation processes are negligible. At intermediate and high proton energies all excitation processes must be taken into account. Up to values on the order of 1 MeV the excitation of single conduction electrons is the prevailing mechanism. However, for very high primary energies the mechanism of core excitation is the dominant one as can be seen from the data in Table 9.1.

The Auger excitation contributes to the yield by directly emitted electrons as well as by the electron cascade. The first contribution which is related to the observed Auger peaks in the energy distributions is considerably smaller than the second one. The Auger excitation processes should be taken into account for quantitative considerations. In particular, at very high primary energies their contribution is comparable to the contributions from the other excitation mechanisms.

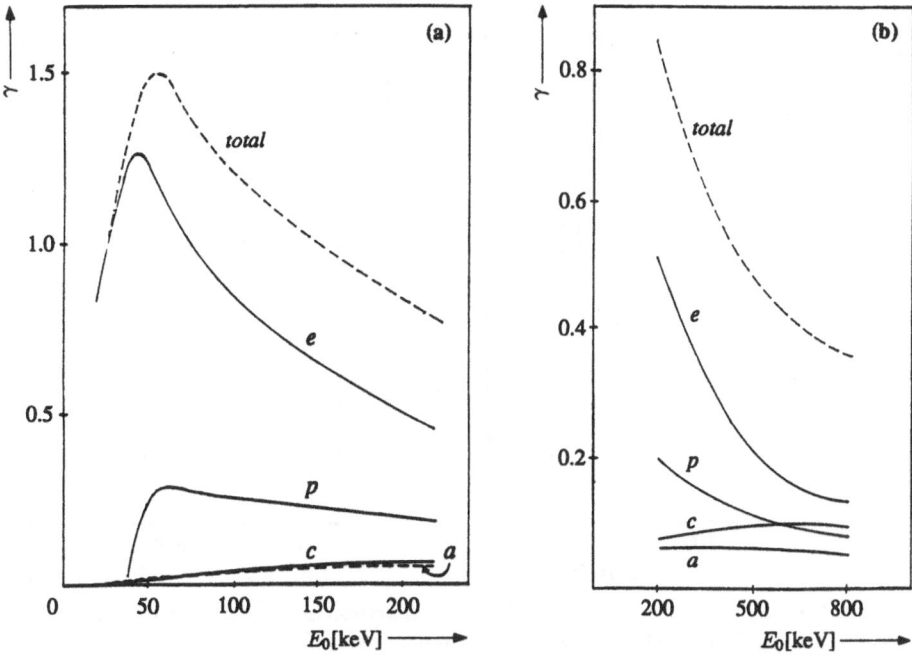

Fig. 9.2. Primary energy dependence of the electron yield γ. Contributions from different excitation mechanisms (see figure caption Fig. 7.9) at (a) low and (b) intermediate and high primary energies (calculated)

Table 9.1. Theoretical and experimental values of electron yields at very high proton energies

E_0	Theory					Expt.
	Contributions to γ				γ	γ
[MeV]	e	p	c	a	total	
5	0.021	0.019	0.037	0.019	0.096	0.20
10	0.011	0.011	0.024	0.013	0.059	0.12

In Fig. 9.3 we compare our theoretical results with the experimental ones mentioned in Chap. 2. In Table 9.1 this comparison is extended to very high proton energies used in the measurements of *Koyama, Shikata*, and *Sakairi* (1981). At low E_0 we obtain agreement with the measured yield values. In particular, the position of the yield maximum at $E_0^{\mathrm{max}} \approx 55\,\mathrm{keV}$ is in accordance with measurements. With increasing E_0 there are deviations from the experimental values. The calculated yield is smaller than the experimental one and differs from the latter by roughly a factor two at high (Fig. 9.3b) and very high (Table 9.1) proton energies.

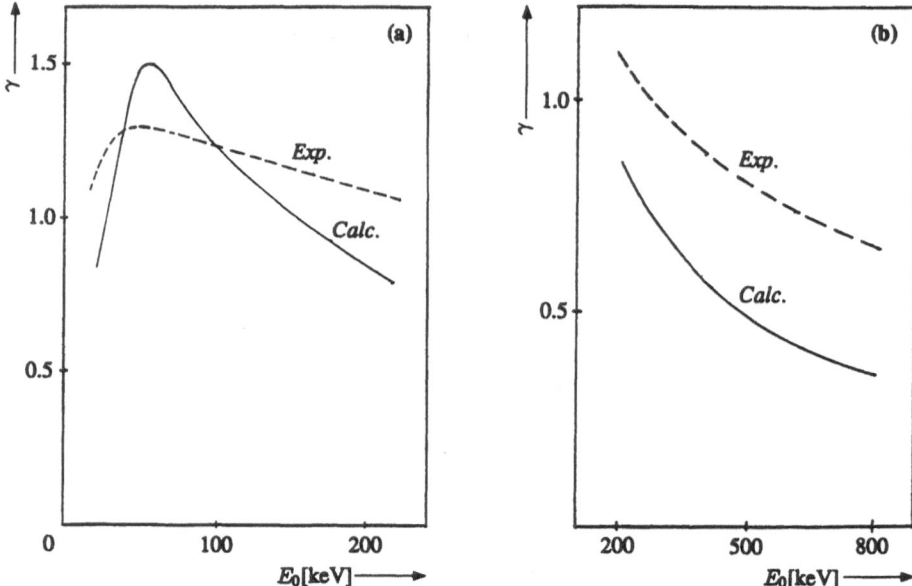

Fig. 9.3. Primary energy dependence of yield γ. Comparison of calculations with experimental results (see Fig. 2.3) at (a) low and (b) high primary energies

9.2 SEE

9.2.1 Energy and Angular Distribution. The energy distribution due to the different excitation mechanisms together with the total energy distribution is shown in Fig. 9.4 for $E_0 = 2\,\text{keV}$. The shape of this total energy distribution, which is confirmed by measurements, is the same as that obtained in the case of IIEE at intermediate and high primary energies. The energy of the cascade maximum and the plasmon shoulder as well as the value of the half-width are common features in the energy distributions of escaping electrons for both particle-induced emission phenomena. From Fig. 9.4 we can see that all excitation mechanisms must be included in the theoretical description of SEE.

For the energy angular distribution of true SE we obtain a cosine distribution for all contributions from different excitation mechanisms and, therefore, for the total angular distribution. This behavior which is obtained for all relevant energies leads to a cosine distribution of the energy integrated angular distribution as observed by *Oppel* and *Jahrreis* (1972).

9.2.2 Yield. As mentioned in Chap. 2 in the case of SEE we consider only the yield δ_p of SE produced by the incident primary beam. In Fig. 9.5 the primary energy dependence of δ_p for the different excitation mechanisms is shown. From this figure we can state that the contribution of core excitation prevails over the other excitation mechanisms in the primary energy range chosen.

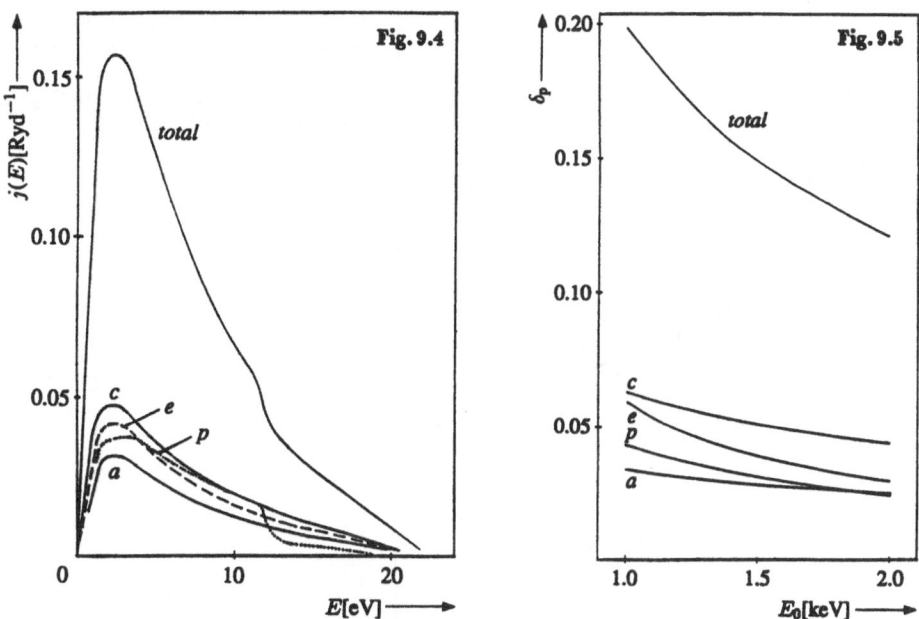

Fig. 9.4. Energy distribution of SE. Contributions from different excitation mechanisms (see figure caption Fig. 7.9). $E_0 = 2\,\mathrm{keV}$ (calculated)

Fig. 9.5. Primary energy dependence of yield δ_p for different excitation mechanisms (calculated)

Table 9.2. Theoretical values of electron yield δ_p at high primary energies

| E_0 [keV] | Contributions to δ_p | | | | | δ_p |
	e	p	c	c (Gryzinski)	a	Total
5	0.012	0.012	0.024	0.018	0.013	0.061
10	0.006	0.006	0.014	0.010	0.008	0.034

Besides the values of E_0 which appear in this figure Table 9.2 presents additional values for higher primary energies which are of interest in scanning electron microscopy. Also for these high primary energies the core excitations supply the greatest contribution to the electron yield.

In Sect. 7.2 we have introduced different models for the evaluation of the excitation rate from core levels. As shown in Fig. 7.17 there are distinct differences between these models with respect to the energy dependence of the excitation function. Nevertheless, we obtain comparable results for δ_p as shown in Fig. 9.6. The values obtained with the atomic model of *Gryzinski* (1965) are somewhat below our calculated values. This is the direct consequence of the smaller number of excited electrons at high excitation energies within the atomic description.

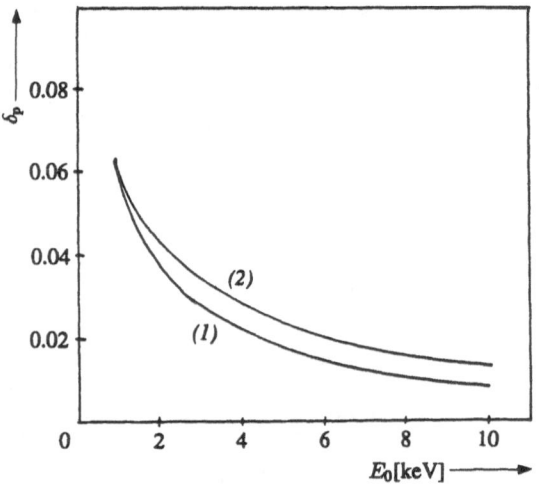

Fig. 9.6. Primary energy dependence of the 2p-core contribution to the yield δ_p. Comparison of the atomic model calculation according to Gryzinski (1965) (*1*) with our results (*2*)

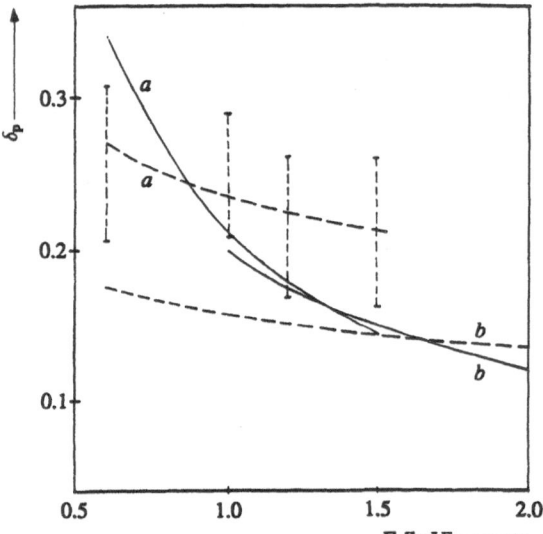

Fig. 9.7. Primary energy dependence of the yield δ_p. Comparison of experimental (*dashed lines*) and theoretical (*solid lines*) results. Experimental values: (*a*) Bindi et al (1980b); (*b*) Bronshtein and Fraiman (1961). Computed values (*a*) Bindi et al (1980b); (*b*) our results

The determination of the yield contribution related to the incident PE δ_p from the experiments is affected by the uncertainty of giving reliable values for the efficiency β of backscattered electrons[2]. Therefore, there are some difficulties in obtaining δ_p with sufficient accuracy. In Fig. 9.7 we compare our calculated values for δ_p with the experimental results given by *Bronshtein* and

---

[2] Different values for β are given in the literature. In the primary energy range of 1 to 2 keV we obtain from experiments values between 4.5 and 8 (Bronshtein and Fraiman 1961; Bindi et al. 1980b). The calculated values are smaller: 2.3, ..., 2.4 (Bindi et al. 1980a) and 2.6 (Reimer 1985).

Fraiman (1961) as well as with the experimental and theoretical results given by *Bindi* et al. (1980b). The inclusion of the Auger contribution leads to an enhancement of the theoretical yield values compared with the results obtained earlier (Rösler and Brauer 1988). In this way we obtain satisfactory agreement between our theoretical yield values and the experimental ones. However, the agreement with the calculated values of *Bindi* et al. (1980b) is more or less accidental because there are distinct differences in the basic assumptions, especially with respect to the contribution of core electrons.

10. The Role of Elastic Scattering

In this chapter the effect of elastic scattering on the energy distribution of emerging electrons and the electron yield will be considered.

As mentioned in Chap. 6 inelastic as well as elastic scattering processes must be taken into account in a realistic description of transport of inner SE. In the following we emphasize the importance of the elastic scattering. This will be the case for both emission phenomena because in both cases the angular distributions of excited electrons are more or less anisotropic. In particular, it is the excitation of single conduction electrons which is strongly anisotropic as shown in Chap. 7. For this type of excitation we demonstrate in Fig. 10.1 the effect of elastic scattering on the angular distribution of the density of inner excited electrons $N(E, \Omega)$.

Neglecting elastic scattering, the angular distribution of $N(E, \Omega)$ shows a decided resemblance to the anisotropic angular distribution of the corresponding excitation function. If we take into account the elastic scattering, we obtain a nearly isotropic distribution. However, it should be noted that the current density of ejected electrons $j(E, \Omega)$ is determined by that part of the internal distribution which is restricted to the escape cone.

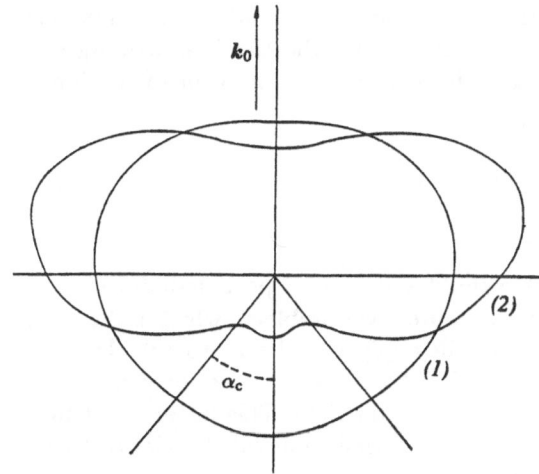

Fig. 10.1. Angular distributions of the density $N(E, \Omega)$ of inner excited electrons for the case of excitation of single conduction electrons including (*1*) and neglecting (*2*) elastic scattering. $E = 10\,\text{eV}$ (measured from the vacuum level). k_0 primary wave vector, α_c aperture of the escape cone (calculated)

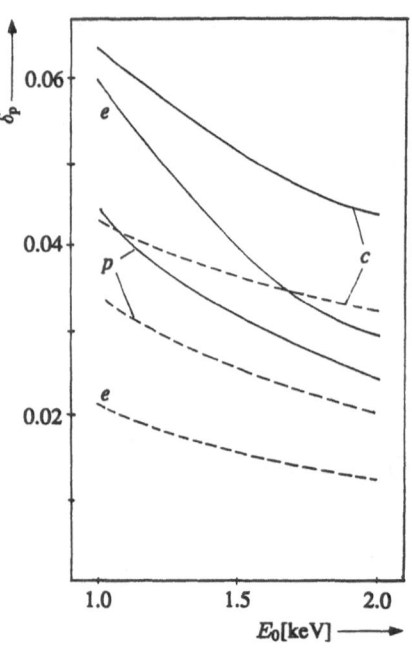

Fig. 10.2. Primary energy dependence of yield δ_p for different excitation mechanisms (see figure caption Fig. 7.9). Effect of elastic scattering. *Solid lines* with, *dashed lines* without elastic scattering (calculated)

It is interesting to consider theoretically the effect of elastic scattering on the measurable quantities, in particular, on the electron yield. The results are shown in Fig. 10.2 for the case of SEE. As can be seen from this figure, for all excitation mechanisms there is an enhancement of the corresponding yield contribution if we take into account the elastic scattering. This is expected because in every case the internal density distribution is enhanced by elastic scattering in the backward direction. For example, in the case of the strongly anisotropic individual excitation of conduction electrons the elastic scattering leads to a considerable rise in the electron yield. We note that the same statements can be obtained in the case of proton-induced electron emission.

An important point in the solution of the transport equation is the influence of the elastic scattering. When taking into account the elastic scattering it is sufficient to restrict oneself to the contributions from $l \leq 2$ in the expansion of $\psi(E, \Omega)$ with respect to Legendre polynomials.

11. Miscellaneous Problems

Starting from the basic concepts for the description of particle-induced electron emission we will consider in this chapter two problems which are closely related to the kinetic IIEE. First, we will apply our theory to calculate the characteristics of electron emission for thin films. Second, we investigate the relationship between stopping power and electron yield. This latter problem is of fundamental interest to the development of semiempirical theories of IIEE.

11.1 Results for Thin Films

Our description of particle-induced electron emission is based on the infinite
slowing-down model, which must be supplemented by the escape conditions.
In this way we obtain the characteristics of the electrons emitted backwards.
However, within such a framework it is also possible to calculate the characteristics of electrons emitted in the forward direction. This possibility enables us
to discuss results for thin films, especially for the forward to backward yield ratio. Representing the current density by an expansion in Legendre polynomials
we have for the electron yield in the forward direction

$$\gamma_f = 2\pi \sum_{l=0}^{\infty} (-1)^l \int_W^{W+50} dE\, l(E) A_l(E) \psi_l(E) . \tag{11.1}$$

The coefficients $\psi_l(E)$ are the same as those which govern the backward yield
(9.2) denoted here by γ_b. Of course, we can expect differences between backward and forward characteristics only for an anisotropic angular distribution of
inner excited electrons. This would be the case in spite of the elastic scattering
which induces an isotropic distribution of inner SE.

In Fig. 11.1 we have plotted the calculated yield ratio γ_f/γ_b for proton impact on Al. As seen in this figure the yield ratio decreases with increasing
primary energy in accordance with other theoretical results (Dubus, Devooght,
and Dehaes 1986). There is no measurement of the yield ratio for proton impact on Al. Experimental results for carbon targets are given by *Meckbach*,
Braunstein, and *Arista* (1975). However, in contrast to the theoretical results
the experimental yield ratio is an increasing function of proton energy below

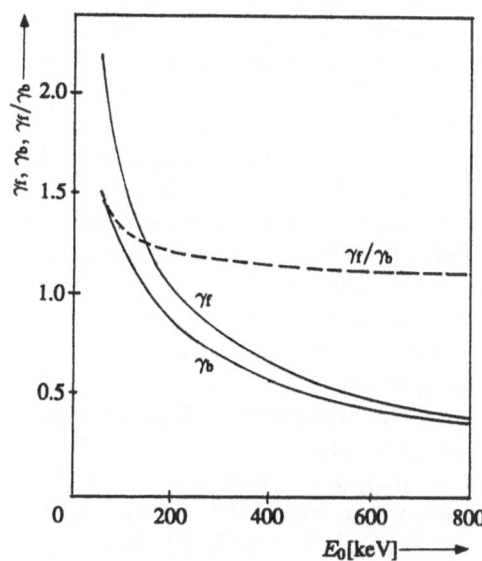

Fig. 11.1. Electron yields (γ_f, γ_b) and forward-backward yield ratio as a function of
the primary energy (calculated)

150 keV. The reason for this discrepancy is not clear at present. One must notice, however, that the backward yield obtained by *Hasselkamp* and *Scharmann* (1983) differs from the mentioned experimental results.

11.2 Stopping Power and Electron Yield in IIEE

In semi-empirical theories of kinetic IIEE one of the basic assumptions is the direct relationship between the number of excited electrons and the stopping power $(-dE_0/dx)$ of the ion. It follows that the electron yield γ is also directly related to the stopping power, i.e.,

$$\gamma = -\Lambda \frac{dE_0}{dx} \ . \tag{11.2}$$

A detailed discussion of these relations has been given (Rösler and Brauer 1984). For Al the empirical relation (11.2) is confirmed by the measurements of *Hasselkamp* et al. (1981) and *Svensson* and *Holmén* (1981). They found $\Lambda \approx (0.11 \text{ to } 0.12) \ 10^{-10} \text{ m/eV}$ in the energy range between 40 and 800 keV.

It is interesting to prove such an empirical relation within a microscopic theory. For this reason we must calculate the stopping power using the same basic assumptions about the different interaction processes as applied in the description of kinetic IIEE. With the transition probability (5.1) the stopping power can be written as

$$-\frac{dE_0}{dx} = \frac{1}{v_0} \sum_{\substack{k_0', k(<k_{\mathrm{F}}) \\ k'(>k_{\mathrm{F}})}} W_{k_0 k \rightarrow k_0' k'}(E_{k_0} - E_{k_0'}) \ . \tag{11.3}$$

Contrary to this simple formula for the stopping power the characteristic quantities for the kinetic IIEE, especially the electron yield, are evaluated within the three-step model (Chap. 3). Therefore, there is obviously no simple relation between stopping power and electron yield (Rösler and Brauer 1989).

The stopping power was calculated, including the energy loss by interaction with single conduction electrons, by excitation of plasmons, and by interaction with core electrons (Brauer and Rösler 1985). It was found that the main contribution to the stopping power in the energy range from 40 up to 800 keV is given by the individual interaction between protons and conduction electrons.

The primary energy dependence of the total energy loss and the electron yield is shown in Fig. 11.2 together with the corresponding experimental results. Experimental values for the stopping power are taken from *Andersen* and *Ziegler* (1977). The calculated stopping power coincides qualitatively with the experimental results. For the position of the maximum (≈ 55 keV) there is complete agreement between theory and experiment.

From the numerical results for the stopping power and electron yield we obtain the ratio $\gamma/(-dE_0/dx)$ which is plotted in Fig. 11.3 together with the experimental values. At low proton energies (E_0 below 100 keV) there is rea-

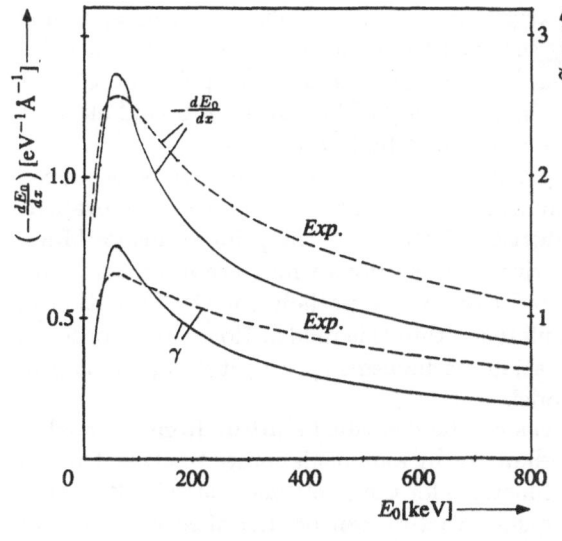

Fig. 11.2. Primary energy dependence of stopping power and electron yield in Al. *Dashed lines*, experiment (Andersen and Ziegler 1977): *solid lines*, theory

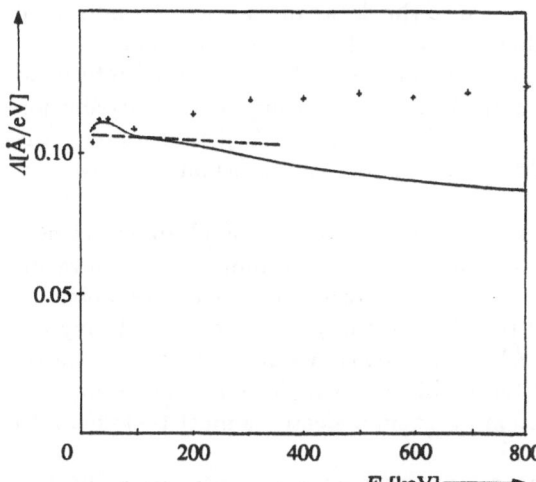

Fig. 11.3. Primary energy dependence of the ratio of electron yield to stopping power. *Solid line* is from theory. The experimental results are given by *crosses* (Hasselkamp et al. 1981) and the *dashed line* (Svensson and Holmén 1981)

sonable agreement between theory and experiment. At intermediate and high proton energies the calculated values are clearly smaller than the experimental ones.

12. Concluding Remarks

In the present work a microscopic theory of particle-induced electron emission for NFE metals is developed. The theory is based on a common description of different processes which govern the excitation of solid state electrons and their transport to the surface of the target. In order to calculate the corresponding scattering rates the dynamically screened Coulomb interaction between point charges is used as a starting point. However, in such a simple picture the effect of projectile electrons as well as the possibility of formation of bound electron states connected with the moving ion cannot be treated.

In a first approximation the system of conduction electrons can be described within the free-electron-gas (jellium) model. This will be sufficient for evaluating the direct excitation of conduction electrons by the primary beam. Moreover, the electron-electron scattering cross sections which are important with respect to the transport of excited electrons, especially for the formation of the electron cascade, can be calculated within this model. However, in order to take into account the other excitation mechanisms (p, c, a) it was necessary to go beyond the simple jellium model.

The excitation from inner shells can be described starting from a model of randomly distributed atoms (randium) or by a more elaborate treatment which takes into account realistic wave functions for the electronic states in the metal. For simple metals, i.e., Al, the band structure can be described in a suitable way within the nearly free electron scheme. The excitation function S_p and the scattering function K_p^s which are related to the decay of plasmons by interband processes can be calculated by using this simple band structure model.

In this context we mention that the inclusion of the K_3-related interband process leads to a distinct, narrow peak in the resulting energy distribution of emerging electrons, in contradiction of experiment. A justification of the restriction to the K_1- and K_2-related interband processes should be given by further investigations.

Some comments on the role of scattering via decay of plasmons is necessary. The corresponding scattering function K_p^s is of minor importance for the number of emerging electrons. However, in order to explain the observed plasmon shoulder in the energy distribution of emitted electrons for heavy ion impact (Fig. 2.4a, Ar^+-impact on Al at $E_0 = 500\,keV$), if the direct excitation of plasmons by the ion is impossible within our simple model of interacting point charges, this scattering function for excited electrons must be taken into account.

Among the different scattering mechanisms we must emphasize the importance of elastic scattering processes. Besides the inelastic electron-electron collisions these processes are responsible for the nearly isotropic angular distribution of internal electrons also in the case of the strongly anisotropic excitation of single conduction electrons. This leads to an increase of the density of inner excited electrons in the backward direction and, therefore, to an enhancement of the number of ejected electrons in comparison with a calculation neglecting elastic scattering.

We have applied our theory to aluminum which is the best investigated NFE metal. For both emission phenomena the theoretical energy distribution of ejected electrons is in accordance with the experimentally determined distributions concerning the position of the cascade maximum, the half-width, and the position of the plasmon shoulder. The electron yield is in qualitative agreement with existing experimental data. In every case the calculated yield values are smaller than the experimental ones at high primary energies.

As an important result of our calculations we can state that in the case of IIEE, below $E_0 \approx 1\,\text{MeV}$ the excitation of single conduction electrons is the prevailing mechanism. At higher primary energies and for the SEE in the range of primary energies considered here (1 to 10 keV) the excitation from core levels is the dominant process. However, in order to obtain reliable quantitative results the other excitation mechanisms must also be taken into account.

Some features of the emissive properties discussed here for Al can generally be observed for other metals, e. g., noble or transition metals. The shape of the energy distribution of ejected electrons is very similar for different types of metals. In particular, the cascade maximum at low energies (at 2 – 3 eV) and the half-width of the distribution (on the order of 10 eV) are common features. Moreover, the electron yield as a function of the primary energy shows the same qualitative behavior for all metals. It is obvious that some aspects of our theory for NFE metals should be valid also for other metals. However, in order to obtain quantitative results it is necessary to take into account their real electronic structure. At present such a program, starting from basic principles, is beyond computational possibilities.

Possible lines of further investigations in the case of IIEE have been mentioned at the beginning of this chapter. In the case of SEE the most important problem is the solution of the transport problem including the effect of scattering processes on the PE. Such effects will be significant at low primary energies if the range of PE is comparable with or smaller than the escape depth. Therefore, to extend the theory of SEE to low primary energies (below 1 keV) the spatial dependence must be taken into account.

Appendix

Historical Survey (1899–1977)

The decade between 1895 and 1905 has proven to be one of the most prolific periods in the development of physics. Besides the discoveries of X-rays, radioactivity, the Zeeman effect, and the electron (Pais 1986), particle-induced electron emission from solids was first disclosed by *Villard* (1899) in Paris, who observed this process when canal rays struck a cathode. Later, in 1902, *Austin* and *Starke* (1902) at the Berlin University were able to see the emission of SE from metals following a bombardment with PE of several keV. They supposed the emission to be enhanced at increasing angle of incidence of the cathode rays; simultaneously they inferred both the disappearance of the emission at normal incidence and the decreasing emission for growing energy of the PE. They also presumed the energy of the ejected electrons to be of the same order of magnitude as that of the incident electrons. Two years later, *Lenard* (1904) was able to correct these statements in two points. First, he ascertained the energy of the SE to be of several eV only; second, his measurement revealed that under normal incidence the emission has no essential difference from the case of oblique incidence. Finally, Lenard observed, for every metal investigated, a maximum of the yield at several hundreds of eV. In 1907 *Laub*, being at that time in the employ of the Würzburg Physical Institute of W. Wien and from 1908 Einstein's first assistant in Bern, confirmed the results of Lenard and undertook the first "attempt to a theoretical interpretation", too. He correctly explained the dependence of the yield on the angle of incidence by assuming that a layer below the surface of the metal is the only source of nearly all SE observed.

After *J. J. Thomson* (1904) in Cambridge and *Rutherford* (1905) in Montreal succeeded in demonstrating the emission of slow electrons following the bombardment of metal plates by α-rays, *Füchtbauer* (1907) carried out a combined investigation of the secondary emission induced by canal and cathode rays. In these two possible excitation mechanisms he found the energy distribution $j(E)$ to be independent of the primary energy E_0, the type of primary radiation, and the angle of incidence. This essentially completed the first stage in the development of SEE before the first World War.

Subsequently a break of nearly two decades ensued, before *Becker* (1925), working at the Heidelberg Physical Institute of Lenard, could continue his earlier research from 1905 on SEE. In 1925 he found the function $j(E)$ to be rather independent of the metal as well as of E_0 for both IIEE by α-rays (Becker 1924) and SEE ($E_0 \approx 1\,\mathrm{keV}$) with the function $j(E)$ showing a maximum at $E \approx 2\,\mathrm{eV}$. Becker (1925) proposed that the energy distribution curve is composed of three parts, from the true SE, the rediffused PE, and the elastically reflected PE. Using his measurements he conjectured the angular dependence of the first to be as in a cosine distribution; simultaneously, he presumed the energy distribution of the SE to be independent of the escape angle.

At that time quantum mechanics came into being, allowing *Fröhlich* (1932) to put forward the first theoretical investigation of the SEE of metals. The extremely complicated treatise pointed mainly to the importance of Bloch bonding of metal electrons in the crystal lattice for the Coulomb interaction with the PE, resulting in a weak dependence of the energy distribution on the primary energy. The possibility of excitation from deep levels was mentioned. Thanks to advances in the quantum theory of solids (Rudberg and Slater 1936) *Wooldridge* (1939) tackled the same problem much more straightforwardly. But in doing so he made serious errors which proved to be very impeding for the discussion of SEE in the following 15 years. This was especially true for the importance of the role of the Fröhlich-Wooldridge processes in the excitation of free and strongly bound electrons. Almost at the same time *Kadyshewich* (1940) finished a classical treatment of the SEE problem using the electron-gas concept of Sommerfeld and taking into account both elastic and inelastic collisions of PE and SE. Thereby a finite emission resulted for the free electron system, too. Thus this theory was, to a certain extent, complementary to that of Fröhlich and Wooldridge. But the employment of a large number of parameters turned out to be very unfavorable for a serious comparison with experimental results.

In the description of the SEE, a much simpler treatment had been developed already previously by *Lukyanov* and *Bernatowitch* (1937), the so-called semi-empirical theory. This theory was based on the assumption

$$\delta(E_0, \vartheta) = A \int_0^R \left(-\frac{dE_0}{ds}\right) e^{-as\cos\vartheta}\,ds$$

for the yield, with ϑ and s denoting the angle of incidence and the straight path length of the PE, respectively. At the same time *Bhawalkar* (1937) investigated independently the case $\vartheta = 0$ (normal incidence), yet incorrectly, using the assumption $E_0^2(s) = E_0^2 - as$ (Whiddington 1912). Thus, he was in principle able to explain correctly the origin of the function $\delta(E_0)$. On the other hand, Lukyanov and Bernatowitch concentrated mainly on the function $\delta(\vartheta)$ expanding it in a series. Not until 1940 did *Salow* formulate the semi-empirical theory very generally and obtain rigorous expressions for $\delta(\vartheta)$ and $\delta(E_0)$. In 1941 *Bethe* carried out a related treatment in a short note referring to the peculiar meaning of electron emission induced by protons. During the years 1940/41 *Kollath* (1947) improved the measurements on the dependence of the energy distribution $j(E)$ of SE both on the primary energy E_0 and the metal examined, simultaneously stating correctly the nearly total independence of $j(E)$ of E_0. He inferred from his results that the SEE has to be interpreted as a material property being solely made visible by the primary beam. The striking similarity of the energy distribution curves for many metals all exhibiting a maximum at about $2\,\text{eV}$ pointed to a fundamental "universality" of the phenomenon. The appearance of the first textbook on SEE by *Bruining* (1942) signified the end of the second stage of the development, the time between the world wars.

The post-war development began with a paper by *Baroody* (1950) in which the author put forward a strongly simplified version of the theory of *Kadyshewich* (1940). Taking into account elastic and inelastic collisions of the SE he presented a clear derivation both of the energy distribution and the yield under the assumption of the existence of free electrons in the metal. Shortly after that *Dekker* and *van der Ziel* (1952) attempted a unified quantum theory of the Baroody and Wooldridge processes; but only *van der Ziel* (1953) was able to describe the latter correctly (Marshall 1952). For the first time he employed in this context a (statically) screened Coulomb potential and concluded the predominance of the Baroody processes (Hachenberg and Brauer 1954, Kollath 1956).

In the same year *Vyatskin* (1953) finished a paper on the SEE of Li, taking the excitation from the K-shell into account. This first quantum-theoretical investigation on the role of strongly bound electrons in SEE led to the conclusion that their contribution to the yield can be considered to be negligible small.

A short time later *Wolff* (1954), in treating the transport process of excited electrons in metals, made an important step by developing the cascade theory. It is this transport process which causes the energy distribution to be independent of E_0 leading to the doubling of the number of electrons at every impact of a SE with an electron of the Fermi sea. Having obtained the solution of the Boltzmann equation where the production rate of excited electrons is the inhomogeneity of the equation, one arrives at statements concerning the quantities $j(E)$ and $\delta_p(E_0)$, provided the mean free path of the electrons is known. Regarding the production rate of excited electrons in alkali metals *Baroody* (1956), using again the pure Coulomb potential, emphasized the dominance of the processes introduced by him in 1950; however, he conceded a non-negligible role to the Fröhlich-Wooldridge processes and the excitation of strongly bound electrons (Dekker 1958).

The first unified theory comprising excitation, transport, as well as escape of free electrons, was developed in 1959 by *Streitwolf* (1959) and *Stolz* (1959). The first author evaluated the energy-angle dependence of the excitation and obtained a strong anisotropy of the excitation function. The second author was able to solve approximately the corresponding Boltzmann equation using inelastic collision cross sections (Wolff 1954). The resulting yield function $\delta_p(E_0)$ was at least one order of magnitude too small, the energy distribution too broad and finally, the angular distribution of the SE strongly deviated from the cosine distribution (Hachenberg and Brauer 1959).

At about the same time *Sternglass* (1957) developed the first semi-empirical theory of the ion-induced electron emission. Concerning the expression for $\gamma(E_0)$ he assumed the excitation ($\sim -dE_0/dx$) to be constant within the escape layer (≈ 50 Å) of the SE and, therefore, obtained an agreement of the E_0-dependences of both quantities. For an appropriately chosen set of parameters a representation was achieved which exhibited fairly good agreement with the measurements available at that time. The yield proved to be dependent only on the velocity and the charge of the ions.

It was not before a further decade had passed that *Fischbeck* (1966) undertook a new attempt towards an understanding of the SEE. In evaluating the excitation from inner shells (of alkali metals) he chose Bloch sums and orthogonalized plane waves for the initial and final states, respectively. The result was a rather isotropic excitation with a relatively low number of energetic electrons being very effective in the cascade process. Therefore, the possibility of an excitation from inner shells could no longer be disregarded (Brauer 1966).

In the same year *Gornyi* (1966) was the first to point to the probable excitation of SE originating from plasmon decay. After having been observed for the first time more than ten years previously (Ruthemann 1941), the phenomenon of plasmon excitation by PE had already been elucidated by *Pines* in 1953. Some time later *Amelio* (1970) refined the theory of Streitwolf and Stolz by using as an approximation the inelastic mean free path $l_{inel}(E)$ derived by Quinn. He thereby succeeded in taking proper account of the plasmon losses during the transport of the SE through the target. But, to be sure, the good agreement between theoretical and experimental $j(E)$-curves was rather accidental.

At the beginning of the 1970s several authors succeeded finally in proving experimentally the existence of SE from the decay of plasmons (Jenkins and Chung 1971; Powell and Woodruff 1972; Everhart et al. 1976). The $j(E)$-curves of highly purified Al surfaces exhibited a structure pointing quantitatively to the plasmon decay. In this context, we should mention the special role of Al as the metal which best fit the free-electron model (Heine 1957). At about the same time *Chung* and *Everhart* (1974) showed that the mean free path $l_{inel}(E)$ derived by Quinn using the dielectric function of *Lindhard* (1954) yields a function $j(E)$ which reproduces the experimental energy distribution much better than the employment of a constant mean free path does. Nevertheless, the authors neglected both the plasmon excitation and the cascade process with the excitation presumed by *Baroody* (1950). In 1977 the same authors (Chung and Everhart 1977) investigating the SEE of Al prompted a new direction of the theory by taking into account the excitation and decay of low-wavelength plasmons. In this way the former Fröhlich-Wooldridge processes were replaced by interband transitions amplified by the plasmon resonance. Besides the excitation processes of free electrons (Thomas-Fermi screening) these transitions should constitute the main portion of the yield. The corresponding transport process was treated in a very simplified way by taking only electron-electron scattering into account. The calculated results regarding the structure of the function $j(E)$ were in good agreement with the experimentally observed one.

Further development of the theory was therefore faced with the task (i) of explaining the role of the excitation from bound states (i. e., especially in Al), (ii) taking into account of the dynamic screening, and (iii) describing the transport process with the help of the Boltzmann equation, including explicitly all scattering processes. In addition, the elastic mean free path $l_{el}(E)$ (Kadyshewich 1940) should probably play an important role in the theory of SEE.

References

Amelio G. F. 1970: J. Vac. Sci. & Technol. 7, 593, Appendix
Andersen H. A., Ziegler J. F. 1977: *Hydrogen Stopping Power and Ranges in All Elements,* (Pergamon, New York). Sect. 11.2
Animalu A. O. E., Heine V. 1965: Philos. Mag. 12, 1249. Sect. 6.1.2
Arendt P. 1969: Phys. Status Solidi 31, 713. Sects. 7.1.3, 7.2.4
Ashcroft N. W., Mermin N. D. 1976: *Solid State Physics* (Holt, Rinehart, and Winston, New York). Sect. 6.1.2
Ashley J. C., Tung C. J., Ritchie R. H. 1979: Surf. Sci. 81, 409. Sect. 6.2.2
Austin L., Starke H. 1902: Ann. Phys. 9, 271. Chap. 1, Appendix
Baragiola R. A., Alonso E. V., Oliva Florio A. 1979: Phys. Rev. B 19, 121. Chap. 2, Sect. 7.1.6
Baragiola R. A., Alonso E. V., Ferron J., Oliva Florio A. 1979: Surf. Sci. 90, 240. Sect. 7.1.6
Baroody E. M. 1950: Phys. Rev. 78, 780. Appendix
Baroody E. M. 1956: Phys. Rev. 101, 1679. Appendix
Becker A. 1924: Ann. Phys. (4) 75, 217. Appendix
Becker A. 1925: Ann. Phys. (4) 78, 228. Appendix
Bennet A. J., Roth L. M. 1972: Phys. Rev. B 5, 4309. Sect. 5.2
Bethe H. A. 1941: Phys. Rev. 59, 940. Appendix
Bhawalkar D. R. 1937: Proc. Indian Soc. 6, 74. Appendix
Bindi R., Lanteri H., Rostaing P. 1980a: J. Phys. D 13, 267. Sect. 5.2

Bindi R., Lanteri H., Rostaing P., Keller P. 1980b: J. Phys. D 13, 2351. Sect. 9.2.2

Bindi R., Lanteri H., Rostaing P. 1987: Scanning Microscopy 1, 1475. Chap. 1

Brauer W. 1966: *Einführung in die Elektronentheorie der Metalle* (Akademische Verlagsgesellschaft Geest und Portig, Leipzig). Appendix

Brauer W., Rösler M. 1985: Phys. Status Solidi (b) 131, 177. Sect. 11.2

Bronshtein I. M., Fraiman B. S. 1961: Sov. Phys. – Solid State 3, 1188. Sect. 9.2.2

Bruining H. 1942: *Die Sekundär-Elektronen-Emission fester Körper* (Springer, Berlin). Appendix

Chung M. S., Everhart T. E. 1974: J. Appl. Phys. 45, 707. Appendix

Chung M. S., Everhart T. E. 1977: Phys. Rev. B 15, 4699. Sects. 7.2.1, 8.2, Appendix

Dekker A. J., van der Ziel A. 1952: Phys. Rev. 86, 755. Appendix

Dekker A. J. 1958: "Secondary Electron Emission", in *Solid State Physics*, Vol. 6, 251 (Academic, New York, London). Appendix

Devooght J., Dubus A., Dehaes J. C. 1987: Phys. Rev. B 36, 5093. Chap. 3

Dubus A., Devooght J., Dehaes J. C. 1986: Nucl. Instrum. & Methods B 13, 623. Chap. 3, Sect. 11.1

Dubus A., Devooght J., Dehaes J. C. 1987: Phys. Rev. B 36, 5110. Sect. 5.2

Dubus A., Devooght J., Dehaes J. C. 1989: preprint (Université Libre de Bruxelles, Service de Metrologie Nucleaire). Chap. 3

Echenique P. M., Flores F., Ritchie R. H. 1988: Nucl. Instrum. & Methods B 33, 91. Sect. 7.1.6

Everhart T. E., Saeki N., Shimizu R., Koshikawa T. 1976: J. Appl. Phys. 47, 2941. Chap. 2, Appendix

Fischbeck H. J. 1966: Phys. Status Solidi 15, 387. Sects. 7.1.3, 7.2.4, Appendix

Fitting H. J. 1974: Phys. Status Solidi (a) 26, 525. Chap. 2

Fitting H. J., Reinhardt J. 1985: Phys. Status Solidi (a) 88, 245. Chap. 3

Fröhlich H. 1932: Ann. Phys. (5) 13, 229. Appendix

Füchtbauer Ch. 1907: Ann. Phys. 23, 301. Appendix

Ganachaud J. P., Cailler M. 1979: Surf. Sci. 83, 498. Chap. 3, Sect. 6.1.1

Gornyi N. B. 1966: Fiz. Tverd. Tela 8, 1939. Appendix

Gryzinski M. 1965: Phys. Rev. 38, 305. Sects. 7.2.4, 9.2.2

Guinea F., Flores F., Echenique P. M. 1982: Phys. Rev. B 25, 6109. Sect. 7.1.6

Hachenberg O., Brauer W. 1954: Fortschr. Phys. 1, 439. Appendix

Hachenberg O., Brauer W. 1959: "Secondary Electron Emission from Solids", in *Adv. Electronics. Electron Phys.*, Vol. XI, 413 (Academic, New York, London). Appendix

Hasselkamp D. 1985: "Die ioneninduzierte kinetische Elektronenemission von Metallen bei mittleren und grossen Projektilenergien"; Habilitationsschrift, Universität Giessen. Chap. 1

Hasselkamp D. 1988: Comments At. & Mol. Phys. 21, 241. Chap. 1

Hasselkamp D., Lang K. G., Scharmann A., Stiller N. 1981: Nucl. Instrum. & Methods 180, 349. Chap. 2, Sect. 11.2

Hasselkamp D., Scharmann A. 1982: Surf. Sci. 119, L388. Chap. 2

Hasselkamp D., Scharmann A. 1983: Vak.-Tech. 32, 9. Sect. 11.1

Heine V. 1957: Proc. Roy. Soc. A 240, 340. Appendix

Herman F., Skillman S. 1963: *Atomic Structure Calculations* (Prentice-Hall, Englewood Cliffs New Jersey). Sect. 7.1.3

Jenkins L. H., Chung M. F. 1971: Surf. Sci. 28, 409. Appendix

Kadyshevich A. E. 1940: J. Phys. (USSR) 11, 115. Appendix

Kollath R. 1947: Ann. Phys. (6) 1, 257. Appendix

Kollath R. 1956: "Sekundärelektronen-Emission fester Körper bei Bestrahlung mit Elektronen", in *Encyclopedia Phys.* Vol. XXI, 232. Appendix

Koshikawa T., Shimizu R. 1974: J. Phys. D 7, 1303. Chap. 3

Koyama A., Shikata T., Sakairi H. 1981: Jpn. J. Appl. Phys. 20, 65. Chap. 2

Krane K. J. 1978: J. Phys. F: Metal Phys. 10, 2133. Sect. 5.1

Laub J. 1907: Ann. Phys. 23, 285. Appendix

Lenard P. 1904: Ann. Phys. 15, 485. Appendix

Levinson H. J., Greuther F., Plummer E. W. 1983: Phys. Rev. B 27, 727. Sect. 7.1.2

Lindhard J. 1954: K. Dan. Vidensk. Selsk. Mat.-Fys. Medd. 18, 1. Sect. 5.1. Appendix
Lukyanov S., Bernatowich W. N. 1937: Zh. Eksp. & Teor. Fiz. 7, 856. Appendix
Lyo S., Joy D. C. 1988: Scanning Microscopy 2, 1901. Chap. 3
Meckbach W., Braunstein G., Arista N. 1975: J. Phys. B 8, L344. Sect. 11.1
Manson S. T. 1972: Phys. Rev. A 6, 1013. Sect. 7.2.3
Marshall T. F. 1952: Phys. Rev. 88, 416. Appendix
Oppel W., Jahrreis H. 1972: Z. Phys. 252, 107. Chap. 2, Sect. 9.2.1
Paasch G. 1969: Phys. Status Solidi 38, K123. Sect. 5.1
Pais A. 1986: *Inward Bound*, (Clarendon Press, Oxford). Appendix
Pendry J. B. 1980: private communication. Sect. 6.1.1
Penn D. R. 1976: Phys. Rev. B 13, 5248. Sect. 6.2.2
Pillon J., Roptin D., Cailler M 1976: Surf. Sci. 59, 741. Chap. 3
Pines D. 1953: Phys. Rev. 92, 626. Appendix
Pines D. 1963: *Elementary Excitations in Solids* (Benjamin, New York). Sect. 5.1
Powell B. D., Woodruff D. P. 1972: Surf. Sci. 33, 437. Appendix
Puff H. 1964: Phys. Status Solidi 4, 125. Chap. 3
Quinn J. J. 1962: Phys. Rev. 126, 1453. Sect. 6.2.2, Appendix
Raether H. 1980: *Excitation of Plasmons and Interband Transitions by Electrons*, Springer Tracts in Modern Physics, Vol. 88 (Springer, Berlin, Heidelberg, New York). Sect. 5.1
Reimer L. 1985: "Scanning Electron Microscopy", in *Springer Ser. Opt. Sci.*, Vol. 45, 153. Sect. 9.2.2
Rösler M., Brauer W. 1981a: Phys. Status Solidi (b) 104, 161. Chap. 3, Sect. 6.1.2
Rösler M., Brauer W. 1981b: Phys. Status Solidi (b) 104, 575. Chap. 3, Sects. 6.1.2, 7.1.2
Rösler M., Brauer W. 1984: Phys. Status Solidi (b) 126, 629. Chaps. 2, 3, Sects. 6.1.2, 11.2
Rösler M., Brauer W. 1988: Phys. Status Solidi (b) 148, 213. Chap. 3, Sects. 5.2, 6.1.2, 9.1.2, 9.2.2
Rösler M., Brauer W. 1989: Phys. Status Solidi (b) 156, K85. Sect. 11.2
Rudberg E., Slater J. C. 1936: Phys. Rev. 50, 150. Appendix
Ruthemann G. 1941. Naturwissenschaften 29, 648. Appendix
Rutherford E. 1905: Philos. Mag. 10, 193. Chap. 1, Appendix
Salow H. 1940: Phys. Z. 41, 434. Appendix
Schou J. 1980: Phys. Rev. B 22, 2141. Chap. 3
Schou J. 1987: Scanning Microscopy 2, 607. Chap. 1
Shimizu R., Ichimura S. 1983: Surf. Sci. 133, 250. Chap. 3
Sigmund P., Tougaard S. 1981: "Electron Emission from Solids During Ion Bombardment", in *Inelastic Particle-Surface Collisions*, Springer Ser. Chem. Phys., Vol. 17 (Springer, Berlin–Heidelberg). Chap. 1
Sternglass E. J. 1957: Phys. Rev. 108, 1. Appendix
Stolz H. 1959: Ann. Phys. (7) 3, 197. Appendix
Streitwolf H. W. 1959: Ann. Phys. (7) 3, 183. Sect. 7.2.1, Appendix
Sturm K. 1976: Z. Phys. B 25, 247. Sect. 7.1.2
Sturm K. 1977: Z. Phys. B 28, 1. Sect. 7.1.2
Sturm K. 1982: Adv. Phys. 31, 1. Sects. 5.1, 7.1.2
Svensson B., Holmén G. 1981: J. Appl. Phys. 52, 6928. Chap. 2, Sect. 11.2
Thomson J. J. 1904: Proc. Cambridge Philos. Soc. 13, 49. Chap. 1, Appendix
Tung C. J., Ritchie R. H. 1977: Phys. Rev. B 16, 4302. Sect. 6.1.2
Valkealathi S., Nieminen R. M. 1984: Appl. Phys. A 35, 51. Chap. 3
Villard M. P. 1899: J. Physique 8, 5. Appendix
Vyatskin A. J. 1953: Zh. Eksp. & Teor. Fiz. 24, 429. Appendix
Whiddington R. 1912: Proc. Roy. Soc. 86, 360. Appendix
Wolff P. A. 1954: Phys. Rev. 95, 56. Chap. 3, Appendix
Wooldridge D. E. 1939: Phys. Rev. 56, 562. Appendix
van der Ziel A. 1953: Phys. Rev. 92, 35. Appendix

Theoretical Description of Secondary Electron Emission Induced by Electron or Ion Beams Impinging on Solids

J. Devooght, J.-C. Dehaes, A. Dubus, M. Cailler, and *J.-P. Ganachaud*

With 22 Figures

1. Introduction

When fast charged particles penetrate into a solid, they excite kinetic electrons. Those electrons, which escape from the solid, give rise to an outgoing electron current which can be measured; they are called "Secondary Electrons" (SE). When the incident particles are electrons this phenomenon is called "Secondary Electron Emission" (SEE) and for incident ions it is called "Ion Induced Electron Emission" (IIEE).

The IIEE process can be splitted into kinetic and potential electron emissions. In this contribution, IIEE will mean kinetic electron emission, the other process is discussed by Varga and Winter, in a forthcoming volume of this series. An excellent review on IIEE has been presented by Sigmund and Tougaard (1981).

The IIEE and SEE are very similar phenomena. This similarity, first understood by Bethe (1941), has only been recognized later (Schou 1980). Indeed the incident particle, entering into the solid, undergoes elastic and inelastic collisions. The latter collisions produce kinetic electrons, called "internal secondary electrons", which undergo themselves elastic and inelastic collisions, giving rise to an electron multiplication or cascade process. Some of these electrons escape through the vacuum-medium potential barrier. In the energy range considered in this work, i.e. 100 eV–1 keV electrons and 100–500 keV light ions, the trajectories of the primary electrons are very different from those of the ions. In the thin layer from which SE escape, the energy losses and angular deflections of the ions can be neglected, hence the source of internal SE is spatially uniform. On the contrary, these processes are very important for primary electrons which can experience large angular deflections and energy losses. Therefore, the transport of the primary electrons must be taken into account. These differences are discussed, for instance, by Hasselkamp (1985) and Schou (1988).

Although the primary and internal SE are not distinguishable, the transport problem may be solved in two steps for SEE. First, the primary electron flux is calculated. Second, the internal secondary electron flux is calculated using the primary electron flux in a source term, this step is the same as for incident ions. Most models take advantage of this separation: the primary transport problem is solved separately or even, more simply, the primary electron flux is assumed to be spatially constant.

The following brief historical overview of IIEE and SEE will only emphasize those aspects related to the SE transport models. Other aspects can be found in the contribution by Rösler and Brauer (RB), in this issue.

In the early SEE descriptions of Baroody (1950) and Bruining (1954), the internal SE are produced along the path, assumed to be a straight line, of the primary electron. The escape probability of these electrons was simply given by an exponential law, characterized by a mean escape length. A similar model for IIEE has been proposed by Sternglass (1957). These models are purely phenomenological but they have the advantage to give a simple expression for the outgoing electron current, which can be useful in some instances.

A substantial improvement has been effected by Wolff (1954) who included the cascade process in his SEE model which is the first application of the "infinite medium slowing down" model to SE transport. This model has been improved by Stolz (1959) who used the electron excitation function proposed by Streitwolf (1959) instead of the monoenergetic internal SE source chosen by Wolff.

Puff (1964a,b,c) has developed an integral formulation of the SE transport which includes the depth dependence of the internal SE source, the cascade effect and the partial reflection boundary condition. Unfortunately, he did not compare his results to experimental data, at least to our knowledge.

The first realistic SEE models have been proposed in the years 1970. The Puff transport model has been used by Cailler (1969) for studying the SEE from Al and noble metals. He concluded that plasmon decay was giving an important contribution to the SEE from Al and that interband transitions from d-levels was preponderant for noble metals. Bennett and Roth (1972) have studied the influence of the primary electron transport on the SEE yield δ, solving the Boltzmann equation in the form proposed by Bethe, Rose and Smith (1938), to get the primary electron flux. Cailler and Ganachaud (1972) and Ganachaud and Cailler (1973) have studied the SEE from copper using a simplified version of the Puff theory and also a Monte Carlo (MTC) simulation which incorporates the transport of the primary electron. Koshikawa and Shimizu (1974) have also studied the SEE from copper using the MTC method but neglecting the primary electron transport. Chung and Everhart (1977) have used very elaborate collision cross sections in aluminum but their single scattering model is rather crude.

Ganachaud (1977) and Ganachaud and Cailler (1979a,b) have used the MTC method to study SEE from metals, especially aluminum, but also copper and gold. Their interaction model, considered as a standard model in this contribution, will be briefly described in Chap. 5.

Schou (1980) has proposed an interesting model which can be applied to a wide range of incident particles and targets. Starting from the multispecies Boltzmann equation, he succeeded in establishing a simple formula giving the SEE and IIEE yields in terms of known experimental quantities.

Bindi, Lantéri and Rostaing (1980a,b) have solved the Boltzmann equation numerically, the primary electron transport being treated in the continuous slowing down approximation (Lantéri, Bindi and Rostaing 1981; Bindi, Lantéri and Rostaing 1987).

Rösler and Brauer (1981a,b) have used the "infinite medium slowing down" model, which is only valid for a spatially uniform SE source, to calculate the partial yield δ_p for incident electrons on polycrystalline aluminum. They have extended this model to proton induced electron emission (Rösler and Brauer 1984; Brauer and Rösler 1985; Rösler 1987; Rösler and Brauer 1988). Their work is also described in great detail in this issue.

Devooght et al. (1984) and Dubus, Devooght and Dehaes (1986) have proposed two approximate models which are described in Chap. 4. The first model, i.e. the age-diffusion model, gives the Green's function, in the diffusion approximation, in terms of macroscopic quantities derived from the microscopic cross sections. This model involves all the energy, space and time variables. The second model assumes a spatially uniform source, hence it is strictly valid only for IIEE, and gives an estimate of the surface correction to be added to the "infinite medium slowing down" solution.

At last, let us mention the work done by Shimizu and coworkers who used the MTC method to study the electron transport in solids. Although they were mainly interested in the Auger emission (Ichimura and Shimizu 1981), they have also studied SEE (see Ding and Shimizu 1988, for instance).

In Chap. 2, we will give a brief overview on the electron interaction models, emphasizing the aspects not discussed by RB. Different models giving the elastic and inelastic cross sections in aluminum will be compared. We will also discuss some electron interaction models that have been proposed for materials other than aluminum.

In Chap. 3, three MTC methods will be described and compared, emphasizing that only the direct simulation scheme can be used to study SEE. We will also give some additional comments on the statistical aspects of the MTC model, on variance reduction techniques and on the use of vector and parallel computers.

In Chap. 4, we will describe some methods to solve the Boltzmann equation for SE transport. Three approximate models will be discussed: the age-diffusion model, an approximation based upon the integral form of the Boltzmann equation and the transport-albedo model. The purely numerical S_N-multigroup method and the model of Schou (1980,1988) will also be described.

In Chap. 5, we will give some results on the SEE from polycrystalline aluminum and gold. We will put forward the influence of the primary electron transport on the SE yield δ and give some results on the proton induced electron emission.

At last, we will conclude by giving some future prospects.

To help the reader, we give in Table 1.1 a list of the most important quantities used in this contribution.

Table 1.1 Definitions of the most important electron emission characteristics

	Incident electrons
$J(E)$	Outgoing electron current $(\text{eV})^{-1}$
δ	True secondary electron emission yield $\delta = \int_0^{50\text{eV}} J(E)dE$
δ_{p}	True secondary electron yield without primary electron transport
η	Backscattered primary electron emission yield $\eta = \int_{50\text{eV}}^{\infty} J(E)dE$
σ	Total electron emission yield $\sigma = \delta + \eta$

	Incident protons
$J(E)$	Outgoing electron current $(\text{eV})^{-1}$
γ	Total electron emission yield
$\gamma_{\text{F}}, \gamma_{\text{B}}$	Forward and backward emission yields
R_{γ}	$\gamma_{\text{F}}/\gamma_{\text{B}}$

2. Electron Collisions in Solids

A primary beam of electrons impinging on a solid target suffers inelastic and elastic collisions with all the components of the solid. Both types of interactions play an important role in the SEE process. The inelastic collisions bring the electrons of the solid to upper energy levels, giving rise to the so-called cascade effect. The transport of the primary and excited electrons in the solid is very sensitive to the energy and angle dependence of the scattering cross sections. Hence a realistic description of the interaction processes is required to study the SEE.

A priori calculations of the cross sections corresponding to all the processes occurring with a reasonable probability have only been done for nearly-free-electron (NFE) materials, in the frame of the randium-jellium model.

In this chapter, we will give a brief overview on the models used to describe the electron interactions in solids, emphasizing the topics not covered by RB. Especially, we discuss the elastic collision models and some complementary aspects on inelastic interactions in NFE metals like the improvements of the Lindhard dielectric function (1954). At last, we briefly discuss the extensions to other materials.

2.1 Elastic Collisions

Elastic scattering is related to the interactions of the electrons with the real potential $V(r)$ surrounding each ionic core. Various approximations for the potential and different techniques to evaluate the elastic cross section have been proposed. Below, we discuss briefly the cross sections calculated from the first-order Born approximation (FBA) and from the partial wave expansion method (PWEM).

Due to its simplicity, the FBA is often used to calculate the elastic cross section. Although the FBA is expected to fail at low energy (Joachain 1983), it can be used at higher energies instead of the more precise PWEM. The latter method gives the exact differential elastic cross section (see RB in this issue), the potential being incorporated in the phase shifts δ_l. In the low energy range, two methods have been used to calculate the phase shifts: the variable phase approach of Calogero (1967) and the method of Noumerov (1924). Both methods give identical numerical results (Ganachaud 1977). It can be noted that a PWEM including relativistic effects has been used by Ichimura and Shimizu (1981) for Al, Ag, Cu and Au targets. Jablonski (1989) has, however, shown that, at 1 keV, noticeable relativistic effects are only seen at large scattering angles for heavy elements.

The screened Rutherford scattering cross section is an example of the application of the FBA. It is derived from a screened Coulomb potential, which can be written as:

$$V(r) = -\frac{1}{4\pi\varepsilon_0} \frac{Ze^2}{r} \exp\left(-\frac{r}{\varrho_s}\right) \tag{2.1}$$

where Z is the target atomic number, e the electronic charge, ε_0 the permittivity of vacuum and ϱ_s the screening radius. With this potential, the FBA gives the well-known screened Rutherford cross section:

$$\frac{d\sigma_{el}}{d\Omega} = \frac{1}{(4\pi\varepsilon_0)^2} \frac{Z^2 e^4}{4E^2} \frac{1}{(1 + 2\beta - \cos\theta)^2} \tag{2.2}$$

and the total cross section is:

$$\sigma_{el} = \frac{1}{(4\pi\varepsilon_0)^2} \frac{Z^2 e^4}{4E^2} \frac{\pi}{\beta(1 + \beta)} . \tag{2.3}$$

In these relations, E is the electron energy, θ the scattering angle and the parameter β is given as a function of ϱ_s by the relation:

$$\beta = \frac{\hbar^2}{8mE} \varrho_s^{-2} . \tag{2.4}$$

According to the Thomas-Fermi statistical theory of atoms, the radius $\varrho_s = \varrho_{TF}$ of the atomic electron cloud that screens the nucleus can be taken as being:

$$\varrho_{TF} = \frac{1}{2}\left(\frac{3\pi}{4}\right)^{2/3} a_0 Z^{-1/3} \cong 0.885 \, a_0 Z^{-1/3} . \tag{2.5}$$

This relation was recently used by Werner and Heydenreich (1984) in a study of the electron transmission and backscattering with a multiple collision model (see Sect. 3.3).

Many authors have used a modified form of this screening parameter. For instance, Tholomier, Vicario and Doghmane (1987) and Tholomier, Doghmane and Vicario (1988)) used the screened Rutherford cross section in the Lenz approximation of the screening radius: $\varrho_s = a_0 Z^{-1/3}$. Shimizu, Ikuta and Murata (1972) and Ichimura and Shimizu (1981) used for the screening parameter β an expression given by Nigam, Sundaresan and Ta-You Wu (1959) which is equivalent to $\varrho_s = \varrho_{TF}/1.12$. Some other expressions were proposed. For instance, Adesida, Shimizu and Everhart (1978) used half the value of the screening parameter as calculated from the above value of ϱ_s and Jousset (1987) took $\beta = 2.61\, Z^{2/3}/E$.

It is worth noting that these forms of screened Rutherford cross sections were used for intermediate or high energy ($E \geq 1$ keV) electron transport but not for SEE calculations.

In the low electron energy range (for SEE calculations), Bindi, Lantéri and Rostaing (1980a) have deduced the screening parameter β from given values of the elastic mean free path $l_{el} = [N_{at}\, \sigma_{el}(E)]^{-1}$, where N_{at} stands for the number of target atoms per unit volume. Hence, for aluminum:

$$\beta(1 + \beta) = 1656.6\, l_{el}(\text{Å})\, (E(\text{eV}))^{-2} . \tag{2.6}$$

Although the Rutherford cross section can be very useful in preliminary calculations, its angular dependence, as given by (2.2), is wrong even for low Z materials (Jablonski 1989).

In order to get a more realistic description of the elastic collisions, two other kinds of potentials were used in earlier works: atomic and solid state potentials. A frequently used atomic potential has been given by Bonham and Strand (1963) for neutral Thomas-Fermi-Dirac atoms, that is:

$$V(r) = -\frac{1}{4\pi\varepsilon_0}\, \frac{Ze^2}{r}\, \sum_{i=1}^{3} \gamma_i\, \exp(-\lambda_i r) \tag{2.7}$$

where

$$\gamma_i = a_i + b_i\, (\ln Z) + c_i\, (\ln Z)^2 + d_i\, (\ln Z)^3 + e_i\, (\ln Z)^4 \tag{2.8}$$

and equivalent expressions for the λ_i's. Values of the constants for the calculation of the γ_i's and λ_i's are found in Bonham and Strand (1963). Such a potential was used, for instance, by Ichimura and Shimizu (1981).

Three solid state potentials have been given for aluminum targets. The first one has been published by Snow (1967) but it has been found by Greisen (1968) that a systematic error exists in the calculations. Ganachaud and Cailler (1979a) used a muffin-tin potential evaluated by Smrcka (1970) by superposing the potentials of the different ionic sites and by using a Slater exchange term. Pendry (1974) has calculated a self-consistent muffin-tin potential using a Hartree-Fock exchange term and has given the phase shifts for Al.

Jousset (1987) (see also Cailler and Ganachaud 1990b) has compared the atomic and Smrcka potentials. He observed dramatic deviations at large radii which induce large differences in the cross section at small scattering angles, especially for low electron energies ($E \leq 100$ eV). Therefore, atomic potentials cannot be used in the study of SEE because of the dominant role played by the low energy electrons.

Ganachaud (1977) compared the elastic mean free path in aluminum calculated with the muffin-tin potentials given by Smrcka (1970) and Pendry (1974), and he observed large differences below 100 eV. However, a similar comparison, made by RB, between the phase-shift values used by Ganachaud and those that they have obtained from Pendry (private communication (1980)) shows a better agreement.

A comparison between the cross sections obtained by using the FBA and the PWEM was performed by Ichimura and Shimizu (1981) and by Jousset (1987). For aluminum, a close agreement was obtained, especially in the high energy range (above several keV). For heavier materials, Ichimura and Shimizu observed that the FBA could no longer give results close to those obtained by the PWEM, the differences becoming larger as the atomic number of the target atom increases. Such restrictions in the use of the FBA are not surprising if considerations on the electron energy and on the width of the potential well, as developed by Schiff (1955), are taken into account (for further details, see Cailler and Ganachaud 1990b).

Ichimura and Shimizu (1981) and Jousset (1987) have also proceeded to a comparison between the cross sections obtained with the two above methods and a screened Rutherford scattering cross section. Ichimura and Shimizu (1981) used $\varrho_s = \varrho_{TF}/1.12$ and observed that the cross sections obtained by the screened Rutherford formula differ from those obtained by the PWEM or the FBA, even for Al. According to Jousset (1987) the total elastic cross section evaluated in Al, with a screening parameter $\beta = 2.61\ Z^{2/3}/E$, differs by 40%, at 1 keV, from the PWEM result. Furthermore, the differential cross sections are very different from those obtained by the PWEM, especially at low energies.

We compare in Fig. 2.1 the differential elastic cross section calculated by PWEM from Smrcka potential to the screened Rutherford cross section with a screening parameter deduced from (2.6). Structures appearing in the PWEM cross section are absent from the monotonous and more isotropic Rutherford cross section.

In Fig. 2.2, we compare the elastic mean free paths in Al calculated by PWEM from the Smrcka potential (1970), the Pendry potential (1974) and the atomic potential given by Bonham and Strand (1963), the mean free path calculated by FBA from the same atomic potential and the screened Rutherford formula using the screening parameters given by Nigam, Sundaresan and Ta-You Wu (1959) and Jousset (1987). An obvious disagreement appears between the screened Rutherford formula and the other calculations. For the atomic potential, the energy reference level is not well defined in the medium and we have not calculated the mean free path below 100 eV.

All these comparisons show that if low energy electrons ($E \leq 100$ eV) have to be considered, a free atom approximation of the potential is not sufficiently

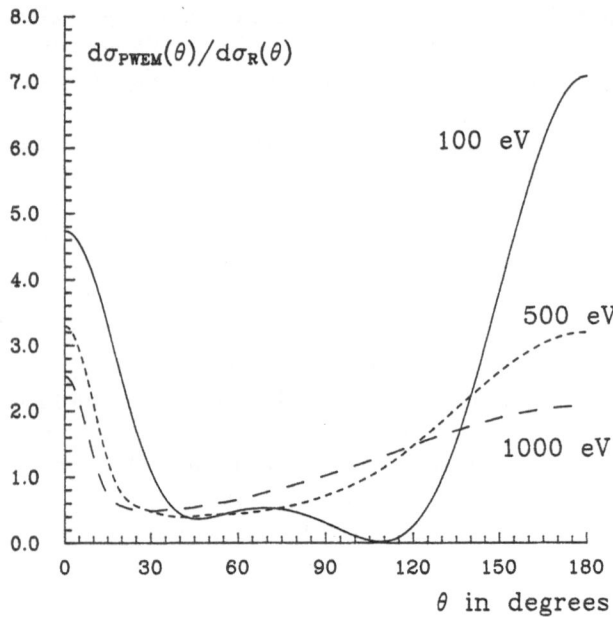

Fig. 2.1. Ratio of the differential elastic collision cross section deduced from the Smrcka potential (1970) by PWEM to the screened Rutherford cross section. The screening parameter β is deduced from the Smrcka cross section using (2.6)

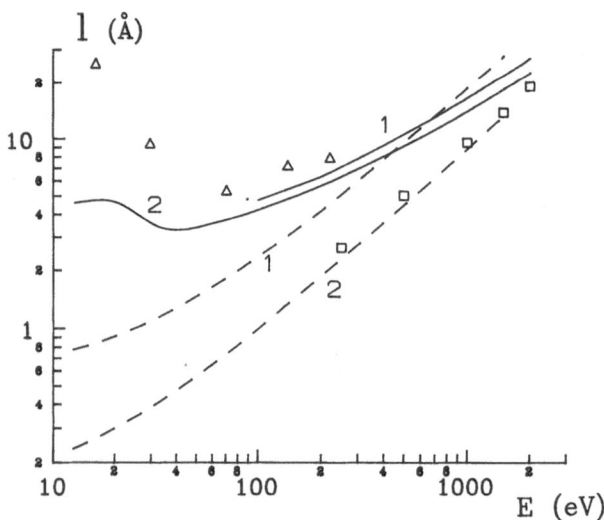

Fig. 2.2. Comparisons of elastic mean free paths. *Solid line* (1): PWEM and the Smrcka potential (1970); *solid line* (2): PWEM and the Bonham and Strand atomic potential (1963); *dashed line* (1): screened Rutherford cross section with a screening parameter taken from Nigam, Sundaresan and Ta-You Wu (1959); *dashed line* (2): screened Rutherford cross section with a screening parameter taken from Jousset (1987); $\triangle$: PWEM and Pendry potential (1974); $\square$: FBA and Bonham and Strand atomic potential

precise and the screened Rutherford cross section is wrong. A better choice lies in a self-consistent potential which takes into account the charge redistribution due to the delocalization of the outer-shell electrons. From this point of view, the spherically symmetric muffin-tin approximation is a very useful form. Evidently, the choice is less important at higher energies. For instance, the atomic potential given by Bonham and Strand (1963) works fine for a study of electron backscattering as shown by Ichimura and Shimizu (1981).

2.2 Inelastic Collisions

A rigorous description of the inelastic interactions in a solid would be rather sophisticated. An essential simplification of this problem is to consider that the solid reacts as a whole to an external charge, which is here an incident electron. The dielectric theory of the response of a solid to a point charge can be regarded as having a sufficiently wide range of validity (see for instance Pines (1963) and Sturm (1982)).

The problem for SEE calculations is that not only information on the inelastic scattering is needed but also information on the excitation of electrons. For this reason, most authors use the Lindhard dielectric function (Lindhard 1954) or an improved version of it. In this context, only NFE metals, such as aluminum, can be studied. A remarkable feature of the Lindhard dielectric function is that it breaks the response function into two parts, the plasmon excitation and the electron-hole pair excitation.

Electron penetration and transmission can be studied in a wider range of materials than SEE because only the scattering cross section is needed. Indeed, a very attractive model for calculating the differential scattering cross section from optical data has been recently proposed by Penn (1987). This model has been used, for instance, by Tanuma, Powell and Penn (1988) and Ding and Shimizu (1989). Unfortunately, this model does not give the electron excitation cross section.

RB have described their model for inelastic electron interactions in NFE materials. In the following, we describe some complementary aspects used in our calculations: improvements of the Lindhard dielectric function, excitation and decay of bulk and surface plasmons and the role played by the inner-shell ionizations.

2.2.1 Dielectric Response Functions for NFE Materials

The Lindhard dielectric function (Lindhard 1954), which is calculated in the random phase approximation (RPA), does not include the short range interactions between the conduction electrons. Hubbard (1957) and Nozières and Pines (1958), for instance, have shown that the RPA is not correct for large momentum transfer. Hubbard (1957) proposed to overcome this problem by introducing a function $G(q)$ which takes into account the depletion of negative charge surrounding an electron. The general principles underlying the static exchange and correlation corrections are discussed for instance by Kugler (1975) and Mahan (1990), the corrected form of the dielectric function being:

$$\varepsilon(q,\omega) = \varepsilon_{\mathrm{L}}(q,\omega) + G(q)\frac{(1 - \varepsilon_{\mathrm{L}}(q,\omega))^2}{1 + G(q)(1 - \varepsilon_{\mathrm{L}}(q,\omega))} \tag{2.9}$$

where $\varepsilon_{\mathrm{L}}(q,\omega)$ is the Lindhard dielectric function.

Many forms have been proposed for $G(q)$. Hubbard (1957) suggested the simple form:

$$G(q) = \frac{1}{2}\frac{q^2}{q^2 + k_{\mathrm{F}}^2} \tag{2.10}$$

where $E_{\mathrm{F}} = \hbar^2 k_{\mathrm{F}}^2/2m$ is the Fermi energy. Vashishta and Singwi (1972) calculated the function $G(q)$ by an iterative procedure and proposed to use the following expression which fits their theoretical results:

$$G(q) = A(1 - \exp(-B(q/k_{\mathrm{F}})^2)) \tag{2.11}$$

where the parameters A and B depend only on the reduced one electron-sphere radius r_s. For aluminum ($r_s = 2.07$), $A = 0.895$ and $B = 0.336$. More recently, complex dynamical exchange and correlation correction functions $\tilde{G}(q,\omega)$ have been used (see for instance Brosens, Lemmens and Devreese (1976)) but they are beyond the scope of this contribution.

Another important modification of the RPA has been introduced by Kliewer and Fuchs (1969) and Mermin (1970). It takes into account the finite lifetime τ of the elementary excitations. The Mermin dielectric functions is given by:

$$\varepsilon_{\mathrm{M}}(q,\omega) = 1 + \frac{[(1 + \mathrm{i}/\omega\tau)\,(\varepsilon_{\mathrm{L}}(q,\omega + \mathrm{i}/\tau) - 1)]}{[1 + (\mathrm{i}/\omega\tau)\,(\varepsilon_{\mathrm{L}}(q,\omega + \mathrm{i}/\tau) - 1)/(\varepsilon_{\mathrm{L}}(q,0) - 1)]} \tag{2.12}$$

From the dielectric functions (2.9) (with $G(q)$ given by (2.11)) and (2.12), we have calculated the inelastic cross sections and compared the SE yields to those obtained using the Lindhard dielectric function. The calculated yields are significantly different (about 20%) as will be shown in Chap. 5.

2.2.2 Bulk and Surface Plasmons

Chung and Everhart (1977) and Ganachaud (1977) have shown that the shoulders observed at about 6 eV and 11 eV in the SE spectrum of aluminum (see Pillon, Roptin and Cailler (1976) and Everhart et al. (1976)), can be explained by the decay of surface and bulk plasmons, respectively, via interband transition of one electron. Surface plasmons are collective excitations that replace bulk plasmons at the vacuum-medium interface (Raether 1988) and can be excited by the primary and secondary electrons crossing the surface. Ganachaud and Cailler (1979a,b) have considered that the interface zone in which surface plasmons can be excited, extends from z_{v} in the vacuum to a depth z_s in the medium, both distances being of the order of 1 Å in aluminum. Deeper in the solid, only bulk plasmons can be excited.

In Chap. 5, we have also taken into account the variation of the surface zone thickness with the energy of the crossing electron (Feibelman 1973).

Bulk plasmons are well-defined collective excitations with an infinite lifetime in the restrictive frame of the Lindhard dielectric theory. In a real solid, bulk plasmons decay due to crystalline structure effects, the presence of impurities and the existence of interfaces (Hasegawa and Watabe 1969; Hasegawa 1971). Ganachaud and Cailler (1979a,b) have assumed that bulk plasmons decay via interband transition of one, two or more electrons from the conduction band.

The damping of surface plasmons has been studied theoretically by Feibelman (1974) and has been incorporated in a SEE model by Ganachaud and Cailler (1979a,b). They have assumed that, like bulk collective excitations, surface plasmons decay via interband transition of one or more electrons from the conduction band.

In Chap. 5, we will assume that bulk and surface plasmons decay by interband transition of only one electron.

2.2.3 Inner-Shell Excitations

In principle, the linear response of the inner-shell electrons to an external perturbation can be included in the dielectric response function of the solid (but it is not incorporated in the expressions given in Sect. 2.2.1). Most authors have used preferentially the classical expressions of the cross sections given by Gryzinski (1965 a,b,c), so avoiding the complexity of the quantum mechanical calculations. RB have made quantum mechanical calculations for inner-shell ionizations, taking Bloch sums to describe the core states and orthogonalized plane waves to describe the excited states. Both quantum mechanical and classical results are discussed by RB.

For further reference, we only present here a comparison between the inelastic mean free paths and stopping powers of the electrons interacting with the jellium and with the L-shell electrons in aluminum (see Table 2.1). While the ionizing collisions are less frequent than the collisions with the jellium, they are the main contribution to the stopping power above 500 eV.

Table 2.1 Comparison of inelastic mean free paths and stopping powers for electron interactions with the jellium (Mermin dielectric function (1970)) and with the L-shell electrons

E (eV)	50	75	100	150	250	500	750	1000
$l_{in}(E)$ (Å) (Jellium)	3.4	3.9	4.5	5.7	8.2	13.8	19.0	24.0
$l_{in}(E)$ (Å) (L-shell)	-	-	276	126	93	103	127	151
$S_e(E)$ (eV/Å) (Jellium)	6.14	5.66	5.06	4.12	2.91	1.64	1.17	0.91
$S_e(E)$ (eV/Å) (L-shell)	-	-	0.31	0.78	1.28	1.44	1.30	1.16

2.3 Extensions to Non-Free-Electron Solids

Even in a normal metal the valence band electrons are only quasi free and must be described as Bloch functions, hence the evaluation of the true dielectric function of the solid can be rather complicated. Therefore, simplifying formulations have been considered as, for instance, that proposed by RB who used the free-electron Lindhard dielectric function in order to calculate the transition probability between Bloch states. Solid state effects have also to be taken into account because they can open new channels as the interband transitions allowing the one electron-hole pair plasmon damping mechanism which would be forbidden in a free electron model. To take these solid state effects into account, RB used different kinds of potentials to calculate the plasmon linewidth.

Different approaches (the simplest being that of Penn (1976a) who has suggested, for instance, to describe the 24 valence electrons of Al_2O_3 as if they were free) have been developed to reach a detailed description of the interactions between energetic electrons and solids (as different as normal metals, noble metals, semiconductors, insulators, compound materials, etc.), some of them being devoted to the description of the SEE properties. Only these latter approaches will be shortly discussed in the next sections. Other topics, as the specific calculation of the dielectric response of materials, will not be addressed.

2.3.1 Dielectric Response Functions of a Model Semiconductor or Insulator

The most pertinent work on this topic was performed by Ritchie and co-authors. We briefly describe here two examples taken from their work. Emerson et al. (1973) studied the electron slowing-down spectrum in silicon, by employing for the valence-electron excitation cross section a model developed by Callaway (1959) and Tosatti and Parravicini (1971). In this model, the dielectric function, derived in the RPA, depends parametrically both on the energy gap and the valence-band electron density. For Al_2O_3, Ritchie et al. (1975) and Tung et al. (1977) used a model related to that employed by Fry (1969) in which the ground-state wavefunction of the valence electrons was described in the tight-binding approximation, while excited states were represented by orthogonalized plane waves.

2.3.2 Use of the Optical Loss Function

The earliest work on the connection between SEE and optical conductibility was performed by Baroody (1956) in the assumption of an unscreened interaction between conduction electrons and primary electrons. This connection was extended to screened interaction between outer-shell electrons and energetic electrons by Cailler (1969) in the scheme of the Boltzmann transport equation. The electron transition probability was deduced from optical measurements and the possibility of the presence of structures in the SE peak of noble metals was then theoretically predicted. For copper, results were obtained by Cailler and Ganachaud (1972) through a Boltzmann equation and a source function derived by the help of the Fröhlich theory (1955) from the optical loss function $\text{Im}[-1/\varepsilon(\omega)]$ measured by Beaglehole (1965). Ganachaud and Cailler (1973) extended their calculations to copper targets in a Monte Carlo simulation model. They used a mean free path and an electron excitation probability function for SE also derived from the measured optical loss function according to:

$$\frac{d\sigma(E, \Delta E)}{d(\Delta E)} = 2\pi C \int_{-1}^{1} d(\cos \theta) \frac{\text{Im}\left(-1/\varepsilon(\Delta E/\hbar)\right)}{2E - \Delta E - 2[E(E - \Delta E)]^{1/2} \cos \theta} \qquad (2.13)$$

and

$$l(E) = \frac{1}{N_{\text{at}}} \left[\int_{E_{\text{th}}}^{E - E_{\text{F}}} \frac{d\sigma(E, \Delta E)}{d(\Delta E)} d(\Delta E) \right]^{-1} \qquad (2.14)$$

where C is a constant which was determined by adjusting the mean free path $l(E)$ to its experimental value and E_{th} is the threshold energy for the interband transitions (for Cu, $E_{\text{th}} = 2$ eV). The model was applied by Mignot (1974), Dejardin-Horgues, Ganachaud and Cailler (1976), Ganachaud (1977) and Pillon et al. (1977). It is worth noting that a similar model using experimental optical data in mean free path calculations was also developed by Powell (1974).

A particular requirement in the study of the response function of rather highly localized states, is to take into account the existence of a local field different from the mean macroscopic field, because of the polarization of these states. This local field can have a substantial influence on the wave vector dependence of the dielectric response function (Nagel and Witten 1975). Hence, it is convenient to estimate correctly the wave vector dependent dielectric function in order to use it in the calculation of electron inelastic mean free paths. Ritchie and Howie (1977) have analyzed the requirements coming from the sum rules for the extension of the "optical" dielectric function to non-zero values of q. Cailler, Ganachaud and Bourdin (1981) have extended the optical loss function to nonzero wave vector values. They used a relation based upon Nagel and Witten and Ritchie and Howie works:

$$\text{Im}\left[-\frac{1}{\varepsilon(q, \omega)}\right] = \alpha_q^2 \, \text{Im}\left[-\frac{1}{\varepsilon(0, \alpha_q\omega)}\right] \qquad (2.15)$$

$$\alpha_q^2 = \frac{1}{1 + bq} \qquad (2.16)$$

where b is a constant to be determined. They used this model to evaluate the electron mean free path in copper.

The use of the optical loss function for calculating the inelastic mean free paths and studying the slowing down of the electrons has now become very popular. Recently, a model of the outer-shell electron excitation was also proposed by Ding and Shimizu (1988) allowing a description of the $E \cdot J(E)$ spectra from Cu, Au and Si. In this model, the experimental optical dielectric function was employed to evaluate the excitation function from the relation:

$$\frac{d\sigma(E, \Delta E)}{d(\Delta E)} = \frac{1}{2\pi a_0} \frac{1}{E} \mathrm{Im} \left[-\frac{1}{\varepsilon(\Delta E/\hbar)} \right] \ln \left(\frac{cE}{\Delta E} \right) \tag{2.17}$$

where c is an adjustable parameter which was chosen in order that the inelastic mean free path, as given by (2.14), fits the experimental data over a wide energy region. This model for the outer-shell electron excitation was combined in a composite code in which (2.17) is used in the low energy cascade region (below 100 eV) and in the near elastic peak region, whereas (2.22) was employed in the remaining intermediate energy region. In this composite code, the contributions of the inner-shells were taken into account through the Gryzinski formula (Gryzinski 1965a,b,c).

Penn (1987) proposed for determining electron inelastic mean free paths in solids a modification of the statistical approximation developed by Tung, Ashley and Ritchie (1979). He assumed that:

$$\mathrm{Im} \left[-\frac{1}{\varepsilon(q, \omega)} \right] = \int \frac{d^3r}{\Omega_{\mathrm{WS}}} \mathrm{Im} \left[-\frac{1}{\varepsilon(q, \omega; r_s^{\mathrm{P}}(r))} \right] \tag{2.18}$$

where the domain of integration is a Wigner-Seitz cell of volume Ω_{WS} and

$$r_s^{\mathrm{P}}(r) = \frac{1}{a_0} \left[\frac{3}{4\pi n_{\mathrm{p}}(r)} \right]^{1/3} \tag{2.19}$$

In this expression, $n_{\mathrm{p}}(r)$ is a pseudo-charge-density chosen to ensure that:

$$\mathrm{Im} \left[-\frac{1}{\varepsilon(0, \omega)} \right] = \mathrm{Im} \left[-\frac{1}{\varepsilon(\omega)} \right] \tag{2.20}$$

where the loss function $\mathrm{Im}[-1/\varepsilon(\omega)]$ is determined from optical or electron-energy loss experiments. Equation (2.20) determines $n_{\mathrm{p}}(r)$. Thus, a knowledge of the optical energy loss function is sufficient to obtain $\mathrm{Im}[-1/\varepsilon(q, \omega)]$. This algorithm was used by Penn (1987), by Tanuma, Powell and Penn (1988) for 31 materials as different as free-electron solids, transition or noble elements and compound materials and by Ding and Shimizu (1989). Its predictions compare fairly well with experimental results.

2.3.3 Extension of the Gryzinski Formulation to the Valence Band

Shimizu and Everhart (1978) have proposed a formulation of the cross section for valence electron excitation from an approximation of the Gryzinski equation:

$$\sigma_{\mathrm{v}}(E) = \frac{1}{3} \frac{\sigma_0}{E_{\mathrm{v}}} \frac{1}{T} \ln \frac{E}{E_{\mathrm{v}}} \tag{2.21}$$

where E_{v} is a mean binding energy of valence electrons chosen so that the total electron stopping power (valence + inner-shell electrons) is equal to the Bethe stopping power ($E_{\mathrm{v}} = 4$ eV for aluminum).

Ding and Shimizu (1988) extended this assumption to the calculation of the energy distribution of SE for Si, Cu and Au by assuming that the Gryzinski formula for the excitation function is also valid for the evaluation of the valence electron excitation. This was written:

$$\frac{d\sigma(E, \Delta E)}{d(\Delta E)} = 4\pi a_0^2 z_{\mathrm{v}} \frac{E_{\mathrm{R}}^2}{\Delta E^3} \left(\frac{E}{E + E_{\mathrm{v}}}\right)^{3/2} \tag{2.22}$$
$$\times \left(1 - \frac{\Delta E}{E}\right)^{E_{\mathrm{v}}/(E_{\mathrm{v}} + \Delta E)} \left\{\frac{\Delta E}{E_{\mathrm{v}}} \left(1 - \frac{E_{\mathrm{v}}}{E}\right) + \frac{4}{3} \ln\left[2.7 + \left(\frac{E - E_{\mathrm{v}}}{E_{\mathrm{v}}}\right)^{1/2}\right]\right\}$$

where E_{R} is the Rydberg energy, z_{v} and E_{v} the occupancy number and the mean binding energy of the valence band, respectively. These two parameters were determined from a fit of the stopping power associated with the valence electron excitations. Though this model is very crude, it gives rather good results for SEE, especially from noble metals where the valence and weakly bound electrons are not distinguished in the excitation function.

2.4 Conclusion

As a conclusion to this chapter, we want to notice that attention must be paid to the potential and the method of calculation used to evaluate the elastic cross section. In fact, because low energy ($E \leq 100$ eV) electrons have to be considered, a description of the potential that is as precise as possible is required and a free atom approximation for the potential could not be sufficiently precise. The first Born approximation has to be used with precaution and likely not for heavy atoms. The screened Rutherford formula does not work well, especially at low energies. A better choice lies in the use of a self-consistent potential and of the PWEM method. However, in such a case, the summation of the phase-shifts has to be performed on a sufficiently large number of partial waves. Otherwise, some accuracy losses can occur in the differential cross section values.

We can also retain the importance of the dynamical screening for the realistic estimation of the individual collision probabilities, as well as the need to include crystalline effects (interband transitions) in the description of the plasmon damping. At last, it can be noticed that important developments in the study of non-free electron materials remain to be made. Up to now, rather elaborate models like the Penn model (1987) have been developed for the calculation of electron scattering cross sections in various materials but there is a serious lack of electron excitation models.

3. Monte Carlo Simulation Models

The Monte Carlo (MTC) method is a statistical sampling technique that was applied from the years 1940 with success to many physical problems. With the development of supercomputers, it has become a standard technique in such areas as particle transport, statistical physics, etc.

For particle transport, the MTC method appears as a natural technique to simulate the physical scattering processes that particles undergo along their path. One can properly talk of "direct simulation" when the sampling is directly made from well-identified single scattering laws (that we hope to represent more or less correctly the true physical processes).

Since the years 1950, MTC methods have been applied to charged particle transport, especially to electron penetration and diffusion problems (see Berger 1963). Its specific application to SEE is more recent. The first simulation codes for SEE have been developed by the groups of Nantes (Cailler and Ganachaud 1972) and Osaka (Koshikawa and Shimizu 1974). Since then, their contribution to this domain remains preponderant (Ganachaud and Cailler 1979b; Cailler and Ganachaud 1990b; Ding and Shimizu 1988), both using the direct simulation model. At electron energies above some tens of keV, two other simulation schemes have been used (Berger 1963; Akkerman and Gibrekhterman 1985): the continuous slowing-down (CSD) and the condensed history or multiple collision schemes.

We will describe the three simulation schemes and the explicit use of MTC simulation for SEE from nearly-free electron metals. We will also discuss the statistics of the results and give some comments on variance reduction techniques and on recent aspects of MTC on vector and parallel computers.

3.1 Direct Simulation Scheme

The basic assumptions of the direct particle transport simulation are the following ones:

– The scattering centers (ionic cores, free electrons) are randomly distributed without any correlation. Hence, any crystalline structure and coherent diffraction effects are neglected. It is worth noting that this assumption is also included in the Boltzmann transport equation (see Chap. 4). However, Kamiya and Shimizu (1976) have taken the electron diffraction effect into account in their MTC simulation.

– The interactions of charged particles with the scattering centers take place locally and instantaneously. Between two successive collisions, the particle propagates freely, keeping its energy and momentum unchanged. Thus, the trajectory of an electron inside the solid can be viewed as a succession of straight line segments connected at the points where the collisions take place and where the energy and angular characteristics of the particle are modified. This assumption is no more valid for an electron in a nonuniform potential. For instance, an electron in the vicinity of a fast ion moving in the solid interacts with the dynamical potential induced by this ion (Müller and Burgdörfer 1990).

In the direct simulation scheme, one follows one particle history and the single scattering events are treated one at a time. The particular application of the direct simulation scheme to the SEE process is described below.

3.2 Continuous Slowing-Down Scheme

This scheme is similar to the Spencer-Lewis form of the Boltzmann transport equation for charged particles (Spencer 1959). The main assumptions are that the energy degradation of the charged particles is a continuous process whereas large angle scatterings occur locally and instantaneously. Such assumptions are valid when a single particle undergoes a large number of collisions in which it loses a small amount of energy without angular deflection and from time to time a large angular deflection without energy loss (for instance, an elastic collision with an ionic core). The complete path of the electron consists in straight line segments. Let the ith segment have a length Δs_i. The residual electron energy E_i after the ith segment is given by:

$$E_i = E_{i-1} - \left| \frac{dE}{ds} \right|_i \Delta s_i \; . \tag{3.1}$$

E_i is the residual electron energy after the ith collision and dE/ds is the stopping power, i.e. the energy loss per unit path length.

The estimations of dE/ds are usually based on the Bethe formula:

$$\frac{dE}{ds} = \sum_k C_k \frac{1}{E} \ln \frac{1.116 \cdot E}{J_k} \tag{3.2}$$

the summation runs over all the inelastic processes, characterized by an ionization potential J_k, occurring in the material. More accurate models can be used for calculating the stopping power dE/ds. Ichimura and Shimizu (1981) and Ding and Shimizu (1988) have used the Gryzinski formula to calculate the partial stopping power due to the ionization of inner-shells while Ding and Shimizu (1988) have used the dielectric theory to calculate the contribution of the valence band for instance.

The large angle deflections incorporate the elastic collisions and sometimes a weaker contribution from inelastic scatterings. For the former ones, a screened Rutherford formula or a partial wave expansion have been used.

The choice of the path length Δs_i is somewhat arbitrary. In practice, one can take a fixed value of the order of magnitude of the total mean free path l_t. The continuous slowing-down scheme needs less computational efforts than the direct simulation scheme but it is important to notice that this model neglects the fluctuations of the energy loss ΔE. It has been shown by Akkerman and Gibrekhterman (1985) that this scheme can only be used in a limited energy range and only for calculating integral characteristics. It is unable to predict differential characteristics as the fine structures appearing in the energy spectrum (especially the elastic peak and its satellites).

3.3 Multiple Collision Scheme

The multiple collision model or condensed history scheme has been used for high energy electrons ($E \geq 10$ keV). The leading idea is based upon the assumption that, in a single collision, the angular deflection is small. A large number of collisions can then be condensed in one multiple collision which takes into account the combined effect of many collisions. As a consequence, the computer time can be significantly reduced.

Akkerman and Gibrekhterman (1985) have studied this method, using the Goudsmit and Saunderson theory to account for the multiple scattering of electrons. They used a path length, between two multiple collisions, which corresponds to about 20 individual collisions and evaluated the energy loss along the electron path from the Landau theory . They concluded that this method has the same drawbacks as the continuous slowing-down model.

A variant of the multiple collision scheme has been proposed by Werner and Heydenreich (1984). The Goudsmit and Saunderson theory provides the multiple scattering solution as a series expansion based on the single scattering law. Werner and Heydenreich proposed to use a simple screened Rutherford cross section for the multiple scattering event, pointing out that such a formula is exact for two limiting cases: a single scattering collision and an infinite number of collisions giving an isotropic distribution. A multiple scattering event presumably corresponds to an intermediate situation. They proposed to connect the screening parameter β to the mean number w of single scattering events along a given path:

$$\beta = C \frac{Z^{2/3}}{E} w^{1.3} \tag{3.3}$$

where C is a constant. The total range of an electron in the solid can be subdivided in a suitable number of segments i, of width Δs_i such that $w_i = \Delta s_i / l_t$, where l_t is the collision mean free path. Werner and Heydenreich applied their model to electron backscattering and transmission for a wide range of energies and targets. They estimated that the above scheme is globally correct in an energy range between 5 and 100 keV.

3.4 Application of Direct MTC Simulation to SEE from Nearly-Free-Electron Materials

The selection of a variable u distributed according to a given probability distribution $f(u)$ is of central interest in Monte Carlo investigations. The usual procedure is to generate a random number R_x uniformly distributed in the [0,1] interval from which u is calculated.

A lot of techniques have been proposed to generate R_x (Knuth 1981; Anderson 1990). The most popular and also the oldest is the mixed congruential method which generates a sequence of integer values by the recurrence relation:

$$x_n = (ax_{n-1} + c) \bmod m \qquad n \gg 0 \tag{3.4}$$

giving $R_x = x_n/m$. The values of a, c, m and x_0 are chosen in order to give the longest sequence. Some of them are known to satisfy a large number of randomness tests (Knuth 1981).

To generate values of u distributed according to $f(u)$, we first generate R_x and compute u by inverting:

$$R_x = F(u) \tag{3.5}$$

where $F(u)$ is the cumulative distribution function. When it is possible the relation (3.5) is inverted analytically, otherwise special techniques like the rejection method must be used (Hammersley and Handscomb 1964).

The application of the direct simulation model to the SEE consists to build the complete history of the incident electron, which is initiated at the surface, and the histories of all the secondary electrons resulting from the cascade. In our model, the semi-infinite solid is assumed to be spatially uniform, except for a narrow surface zone. As a consequence, the interaction probability is constant and the free path of an electron is a random variable distributed according to an exponential distribution. We stop building the history of an electron either when it escapes or when its energy is less than the magnitude of the surface barrier.

A large number of electron histories are sampled in order to minimize the statistical uncertainties.

Let us suppose that the ith collision has occurred at the position $r_i(x_i, y_i, z_i)$ and that the electron moves with an energy E_i in the direction given by the unit vector $\Omega_i(u_i, v_i, w_i)$.

First, the position $r_{i+1} = r_i + L_{i+1}\Omega_i$ is calculated from the sampled value of the free path:

$$L_{i+1} = -l_t(E_i) \ln R_x \tag{3.6}$$

where l_t is the total mean free path.

Next, the nature of interaction must be selected from the probability of occurrence of each interaction:

$$P_k = \frac{l_t(E_i)}{l_k(E_i)} \tag{3.7}$$

where l_k is the partial mean free path corresponding to interaction type k. The next value of R_x is compared successively to P_1, $P_1 + P_2$, ... until $R_x \leq P_1 + \cdots + P_k$, interaction k is then selected.

The final step consists to sample the scattering characteristics, i.e. the energy E_{i+1}, the scattering angle θ and the azimuthal angle ϕ of the scattered electron and eventually of the excited electron. Once θ and ϕ are known, the new direction Ω_{i+1} of the electron can be calculated (see for instance Cashwell and Everett 1959).

To illustrate the sampling procedure, we will only consider two cases: elastic and individual electron-electron collisions in the randium-jellium model.

For an elastic collision, the energy remains unchanged and only the scattering angle must be sampled. Because of the isotropy of the medium, the azimuthal

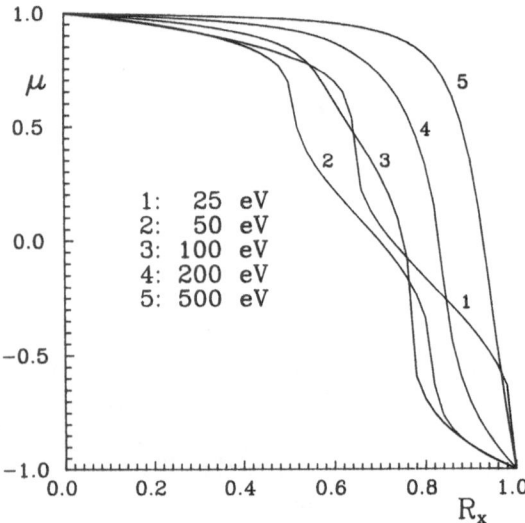

Fig. 3.1. The inverse of the cumulative distribution function $\mu = F^{-1}(R_x)$, from which the elastic scattering angle $\theta = \arccos \mu$ is sampled. The elastic cross section has been calculated from the Smrcka muffin-tin potential in Al

angle ϕ is simply given by $\phi = 2\pi R_x$. The cumulative distribution function corresponding to the scattering angle θ is

$$F(\mu = \cos\theta) = \frac{2\pi \int_\mu^1 d\sigma_{\text{el}}(E_i)/d\Omega' d\mu'}{\sigma_{\text{el}}(E_i)} \qquad (3.8)$$

where $d\sigma_{\text{el}}/d\Omega$ is the differential elastic cross section and σ_{el} is the total elastic scattering cross section. In order to save computer time, we have tabulated the values of μ for several values of $F(\mu)$. It is then easy to invert $F(\mu) = R_x$ by linear interpolation in the table. The inverse function $\mu = F^{-1}(R_x)$ is shown in Fig. 3.1 for the cross section in Al calculated from the Smrcka muffin-tin potential. It is seen that the energy dependence is very smooth.

For a binary electron-electron collision, we must calculate the scattering characteristics of both the scattered and excited electrons. The first step consists to sample the energy loss $\hbar\omega$ and the momentum transfer $\hbar q$, three random numbers being involved.

The probability that an electron of energy E_i loses an energy $\hbar\omega$ by excitation of an electron of the jellium, is proportional to the energy-loss function $\varphi(\omega)$ $(\hbar\omega \leq E)$. The momentum transfer probability for a given energy loss is proportional to the function $\varphi(q|\omega)$ (Ganachaud 1977). The magnitude of the momentum transfer takes values between $q_{\min} = \sqrt{k_{\text{F}}^2 + 2m\omega/\hbar} - k_{\text{F}}$ and $q_{\max} = \sqrt{k_{\text{F}}^2 + 2m\omega/\hbar} + k_{\text{F}}$. Both functions, as calculated from the Lindhard dielectric function, do not depend on the electron energy E, they are shown in Fig. 3.2 and 3.3. In our program, we have tabulated the cumulative distribution functions:

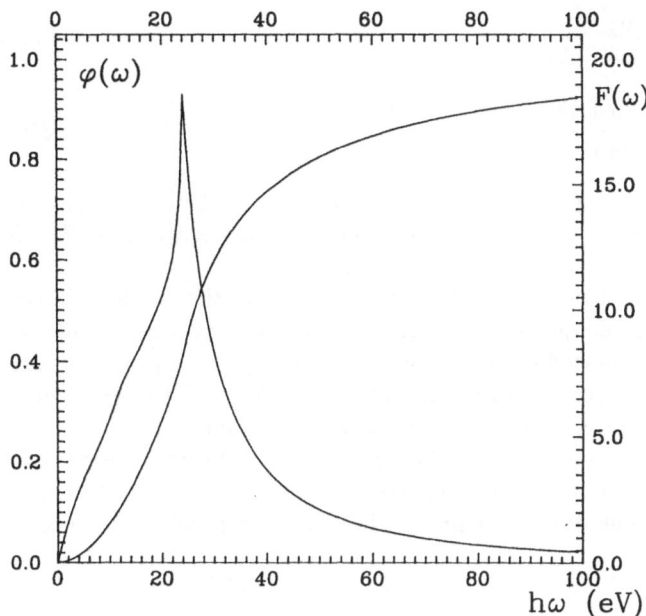

Fig. 3.2. Differential ($\varphi(\omega)$) and cumulative ($F(\omega)$) distribution functions of the energy loss $\hbar\omega$ of an electron in the jellium described by the Lindhard dielectric function

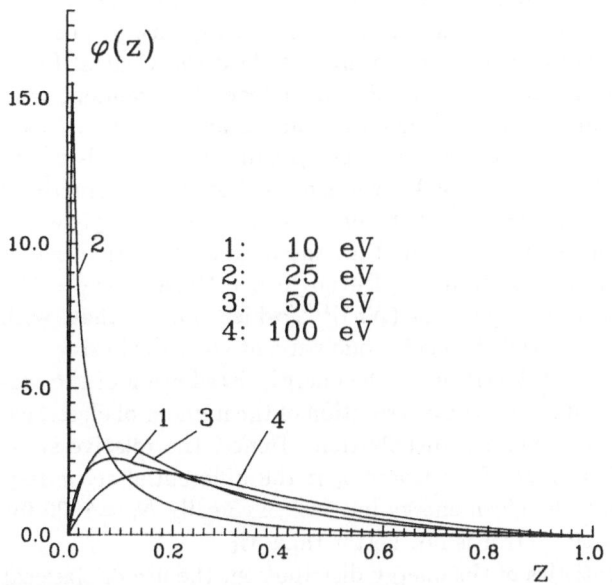

Fig. 3.3. Momentum transfer probability $\varphi(q|\omega)$ as a function of $z = (q-q_{min})/2k_F$, for several energy losses $\hbar\omega$

$$F(\omega) = \frac{\int_0^{\omega} \varphi(\omega')d\omega'}{\int_0^{E_i/\hbar} \varphi(\omega')d\omega'} \tag{3.9}$$

$$F(q|\omega) = \frac{\int_{q_{min}}^{q} \varphi(q'|\omega)dq'}{\int_{q_{min}}^{q_{max}} \varphi(q'|\omega)dq'} . \tag{3.10}$$

It is worth noting that $F(q|\omega)$ is a slowly function of ω (Dubus 1987). Two random numbers are needed to get ω and q, hence the scattering angle θ. The third random number gives the azimuthal angle ϕ.

In the second step, we determine the characteristics of the excited electron whose initial state lies inside the Fermi sphere and final state must be outside this sphere because of the Pauli exclusion principle. The energy and momentum conservation gives the scalar product of the initial momentum and the momentum transfer in terms of ω and q. This implies that the end point of the initial momentum vector lies in a plane perpendicular to q. This condition, together with the Pauli principle, determines an allowed area from which the initial momentum, hence the final momentum, is sampled. Two random numbers are needed in this last step.

3.5 Additional Comments
About the MTC Simulation Method

Assuming that the successive random numbers R_x are uncorrelated, the relative error, i.e. the relative standard deviation, on any estimated quantity must decrease as $1/\sqrt{N}$, where N is the size of the sample. Indeed the Central Limit Theorem states that the mean of a large number of independent random variables, distributed in the same form with mean μ and variance σ^2, tends to a normal variable. The mean of this normal variable is μ and its variance is σ^2/N.

In most cases, the variance σ^2 is not known a priori but it can sometimes be estimated. For example, the statistics of the backscattered electron yield η is close to a binomial distribution if we assume that the only contribution comes from backscattered primary electrons (Dubus, Devooght and Dehaes 1990). The relative standard deviation is then $((1 - \eta)/(N\eta))^{1/2}$, which implies that, with $\eta = 0.2$, about 40000 trials are needed to get a one percent statistical error.

For estimating the standard deviation on the energy distribution of the secondary electrons, we may assume that the distribution of the number of electrons, in a small energy interval, is a Poisson distribution. Hence, the relative standard deviation is $(1/N_i)^{1/2} = (1/\delta_i N)^{1/2}$ where δ_i is the differential secondary electron yield integrated over the given energy interval. Typically $N_i \approx 1000$ for $N = 40000$, hence the statistical error is not better than 3%.

Because of the poor statistics of the energy distribution, the use of classical data smoothing techniques can improve the appearance of an energy spectrum by showing a continuous curve instead of an histogram.

To reduce the computer time, variance reduction techniques can be used. An example of such a technique is the so-called importance sampling. In the specific case of electron emission, the primary electron excites secondary particles up to a depth equal to its range (350 Å for 1 keV electrons in Al targets). The probability that an electron, excited at a depth z by a primary electron, escapes or gives rise to an escaping electron is more or less an exponential distribution which decreases with a characteristic length of about 10 Å. The variance can be reduced by sampling more electrons close to the surface than deeper in the solid. Instead of using the true depth distribution of excited electrons $f(z)$, we use a biased distribution $f^*(z)$ taking into account the escape probability of an electron created at a depth z and give a weight $f(z)/f^*(z)$ to the cascade resulting from this electron.

Many other variance reduction techniques exist, see for instance Mac Grath and Irving (1974).

At last, we can take advantage of the vector and parallel architectures of the modern computers. The conventional history-based simulation of particle transport, in which one follows the history of one particle at a time, can not be vectorized. Recent progress has shown that significant gains in performance can be attained by totally restructuring the conventional Monte Carlo algorithm to be compatible with the vector architecture. The event-based approach, in which electrons histories are splitted into events, has been demonstrated to be very efficient, resulting in excellent vectorization and impressive speedups with respect to the conventional history-based approach (see Brown and Martin 1985). Computer time can be reduced by a factor up to 20 when using this scheme (for computers with 64 word vectors). When using a parallel computer, the computer time can be reduced by another factor approximatively equal to the number of processors in parallel.

We plan to implement variance reduction techniques and the event-based algorithm in our simulation code.

4. Transport Models

Attempts to solve the Boltzmann equation to study electron transport were made in the late thirties, at about the same time as similar attempts for neutron transport. Although a tremendous amount of work, related to nuclear reactor development, has yielded highly sophisticated methods for the study of neutron and, in the same context (fusion and fission), of charged particles (electrons, ions) or photon transport, little of this work has been used for low energy SE transport. It is true that the sheer complexity of the problem, which may involve up to seven independent variables (3 for space, 3 for velocity and 1 for time), leaves little room for general purpose methods and the key to success is the careful use of the simplifications allowed by the problem investigated.

Methods may be classified in three groups:

1. analytical

2. mixed analytical-numerical

3. purely numerical.

In the last category we find methods like Monte Carlo simulation, discussed in Chap. 3, which may avoid, in principle, any simplifying assumption and is therefore invaluable to test hypothesis and provide realistic results. However, it is very time consuming and does not give the same insight on physical relations as analytical models do. None of these, however, can give reliable results on all aspects of a problem simultaneously and mixed analytical-numerical methods provide a good compromise.

The solution of the Boltzmann equation is the angular flux $\phi(r, v, t)$ or $\phi(r, E, \Omega, t)$. Some methods give directly the angular flux. However, the limited possibilities of experimental verification have suggested methods which yield, for instance, the angle and time integrated energy distribution at the surface of a half-space. Therefore a further classification involves "invariant imbedding" methods adapted to the latter case as distinct from general purpose methods (Chandrasekhar 1960; Bellman, Kalaba and Prestrud 1962).

The key input to the Boltzmann equation, beside the geometry, is the set of differential cross sections which are, if realistic, quite involved and virtually intractable for analytical treatment. However two facts must be noticed:

– Boltzmann equation integrates these cross sections multiplied by angular fluxes over angle and energy. A further integration over space results from the transport process itself. The exit spectrum has a final angular smoothing due to the boundary condition (4.4). The end result is therefore quite insensitive to detailed assumptions and this explains, partly, why inadequate models have survived so long the experimental tests.

– Model cross sections like power-laws of energy have been found suitable for analytical treatment (like Mellin transform) (Williams 1979) or for general orientation studies in electron and ion transport (Lux and Pázsit 1981). This does not mean that the cross section modelling has little influence on the choice, or the performance of the approximate solution to the Boltzmann equation: quite the contrary is evidenced by the late recognition (Ganachaud and Cailler 1979a,b) that elastic scattering was dominant for $E \ll 100$ eV electrons with the ensuing quasi isotropy of the flux.

Most methods sacrifice a few independent variables, by means of simplifying assumptions, and concentrate the bulk of the work on the key variables of the problem investigated. For example, sophisticated methods like Wiener-Hopf allow exact solutions for half-spaces at the cost of drastic and irrealistic assumptions on the cross sections (Williams 1971; Duderstadt and Martin 1979).

The most reasonable assumption is the one related to a homogeneous plane half-space medium. Homogeneity is valid if irregularities of density or composition have a characteristic length smaller than the scattering mean free path. Little attention seems to be paid to surface irregularities and possible effects are probably averaged out.

The influence of the boundary is shown in two effects:

– the boundary conditions must allow for an escape cone due to the workfunction, a feature inexistent for neutral particles;

– the angular flux is influenced by the fact that no electron enters the medium from the vacuum half space.

Many treatments ignore the second requirement and are therefore not true space dependent models. A standard result in linear transport theory is the fact that semi-infinite medium problems can be reduced to infinite medium problems ("Placzek lemma") (Case, de Hoffman and Placzek 1953; Case and Zweifel 1967) provided a fictitious angular-energy dependent source is placed at the interface, which gives rise to a spatial transport transient. Since observed effects are dependent on the electron distribution in the "escape depth", it is important to assess properly the boundary condition at the surface where the transient appears.

The angular variable plays usually a great role in high energy ($\geq$ 10 keV) electron transport where strong anisotropy of the cross section is reflected on the angular flux. Spencer-Lewis or Fokker-Planck methods allow for anisotropy and numerical methods like S_N (O'Dell and Alcouffe 1987; Lewis and Miller 1984) must use tailored angular quadratures weighted in the forward direction. Secondary electron transport in the eV range is on the contrary fairly isotropic and the standard "diffusion" (or P_1) distribution linear in the angle cosine is generally valid, owing to the importance of the elastic scattering.

The time variable is completely ignored, except in one case (Devooght, Dubus and Dehaes 1987a), either because problems are stationary or because time dependence is unobservable directly. However, it appears indirectly for transient electric field problems in beam-foil experiments (Gay and Berry 1979; Dehaes, Carmeliet and Berry 1989).

The most important variable is by far the energy of the electron. Despite the filtering introduced by the transport process itself some conspicuous features of the cross sections, like plasmon decay, may still appear in the exit energy spectrum of SE. Experience has been gained in the field of neutron transport with "synthetic kernels" (Williams 1966) which are fictitious cross sections which share with real cross sections some essential features, and give rise to differential equations of first or second order in some energy related variable. Synthetic kernels need not "resemble" the true cross sections, nor even be positive everywhere but moment matching is an essential feature: the larger the number of moments conserved, the better the result. Moreover they provide a systematic way to bridge the gap between the complexity of the cross sections and the smoothness required for the efficient solution of the Boltzmann equation.

The "age" approximation of Fermi belongs to that category. Although discredited for high energy electron transport because it is incompatible with strong angular anisotropy it is quite acceptable for SE transport (Devooght, Dubus and Dehaes 1987a).

There is an essential feature of the energy variable which appears only for low energy electrons, i.e. the cascade process resulting from extraction of electrons from the Fermi sphere. The analogue process for neutron transport is the fission process, much simpler because of its isotropy. However for low energy primary electrons, there is no possibility to distinguish between the two families of primary and secondary electrons. The collision kernel for SE transport is then made of two terms: the "true" scattering term and the excitation term.

In the following section, we will briefly discuss the form of the Boltzmann equation appropriate to SE transport and the partial reflection boundary condition at the vacuum-medium interface. The primary electron transport will not be considered in this chapter, though it must be taken into account to describe completely the SEE process (see Chap. 5). Then we will describe an approximate model for solving the Boltzmann equation which is appropriate to low energy (10–100 eV) electron transport, i.e. the "age-diffusion" model. The integral form of the Boltzmann transport equation will also be discussed as well as two models giving the SE current for a spatially uniform source. The first one is based on an approximate solution of the integral equation and the second one an approximation of the surface Green's function, i.e. the "transport-albedo" model. We also give some comments on the numerical solution of the Boltzmann equation and on the Schou model.

4.1 The Boltzmann Equation

Our starting point will be the linearized Boltzmann equation involving only the SE (see Dubus, Devooght and Dehaes (1990)), the transport of the primary particle (electron or ion) being considered apart and incorporated in the Boltzmann equation as a source term. The medium is assumed to be spatially uniform and isotropic. Moreover any plasmon transport is ignored: the plasmons are assumed to decay in electron-hole pairs virtually at the place of their creation. Then the transport equation reads

$$\left(\frac{1}{v} \frac{\partial}{\partial t} + \boldsymbol{\Omega} \cdot \boldsymbol{\nabla} + \Sigma_s(E) \right) \phi(\boldsymbol{r}, E, \boldsymbol{\Omega}, t) \tag{4.1}$$
$$= \int \int \hat{\Sigma}_s(E' \to E, \boldsymbol{\Omega}' \to \boldsymbol{\Omega}) \phi(\boldsymbol{r}, E', \boldsymbol{\Omega}', t) dE' d\boldsymbol{\Omega}' + Q(\boldsymbol{r}, E, \boldsymbol{\Omega}, t)$$

where ϕ is the electron flux and Q is the independent source of electrons excited by the primary particle along its path.

The scattering kernel $\hat{\Sigma}_s(E' \to E, \boldsymbol{\Omega}' \to \boldsymbol{\Omega})$ is the sum of the scattering cross section $\Sigma_s(E' \to E, \boldsymbol{\Omega}' \to \boldsymbol{\Omega})$ and the excitation cross section $\Sigma_s^s(E' \to E, \boldsymbol{\Omega}' \to \boldsymbol{\Omega})$. The presence of the latter cross section in the Boltzmann equation is responsible for the electron multiplication, the inelastically scattered electron and the excited one both taking part to the electron cascade. The cross section $\Sigma_s(E)$ is the total scattering cross section:

$$\Sigma_s(E) = \int \int \Sigma_s(E \to E', \boldsymbol{\Omega} \to \boldsymbol{\Omega}')dE'd\boldsymbol{\Omega}' . \tag{4.2}$$

The escape process is usually described by a constant potential barrier at the vacuum-medium interface whose amplitude is $U_0 = E_F + \Phi$ where Φ is the work function. For an electron of energy E in the solid, the escape condition is

$$|\boldsymbol{\Omega} \cdot \mathbf{1}_z| = |\mu| = |\cos\theta| \geq \mu_c = \sqrt{\frac{U_0}{E}} \tag{4.3}$$

where θ is the incidence angle measured with respect to the interior normal to the surface which is also the z-axis (see Fig. 4.1). When $|\mu| \leq \mu_c$ the electron is specularly reflected. These conditions give rise to the following ingoing flux boundary condition

$$[\phi(\boldsymbol{r}, E, \boldsymbol{\Omega}, t)]_{z=0} = H(\mu_c(E) - \mu)[\phi(\boldsymbol{r}, E, \boldsymbol{\Omega} - 2\mu\mathbf{1}_z, t)]_{z=0} \quad \mu \geq 0 \tag{4.4}$$

where H is the Heaviside step function.

In many instances, we are not interested in the full spatial dependence of the electron flux. Because of the isotropy of the medium, we can write the transport equation in plane geometry:

$$\left(\frac{1}{v}\frac{\partial}{\partial t} + \mu\frac{\partial}{\partial z} + \Sigma_s(E)\right)\phi(z, E, \mu, t) \tag{4.5}$$

$$= \int_{-1}^{1}\int_{E}^{\infty} \hat{\Sigma}_s(E' \to E, \mu' \to \mu)\phi(z, E', \mu', t)dE'd\mu' + Q(z, E, \mu, t)$$

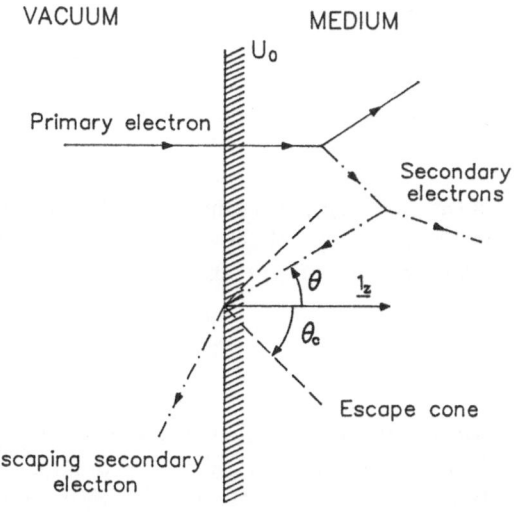

Fig. 4.1. Sketch of the secondary electron emission process. Only electrons in the escape cone can escape from the solid

with the partial reflection boundary condition similar to (4.4):

$$\phi(0, E, \mu, t) = \begin{cases} 0 & \mu_c \le \mu \le 1 \\ \phi(0, E, -\mu, t) & 0 \le \mu \le \mu_c . \end{cases} \tag{4.6}$$

Many methods for solving (4.5) involve the expansion of the angular flux into Legendre polynomials. In the next sections, the following expansions will be used:

$$\phi(z, E, \mu, t) = \sum_{l=0}^{\infty} \frac{2l+1}{2} \phi_l(z, E, t) P_l(\mu) \tag{4.7}$$

$$Q(z, E, \mu, t) = \sum_{l=0}^{\infty} \frac{2l+1}{2} Q_l(z, E, t) P_l(\mu) \tag{4.8}$$

$$\hat{\Sigma}_s(E' \to E, \mu' \to \mu) = \sum_{l=0}^{\infty} \frac{2l+1}{2} B_l(E' \to E) P_l(\mu') P_l(\mu') . \tag{4.9}$$

These expansions are especially useful when we consider the secondary emission induced by fast ions which produce a stationary and spatially uniform source of electrons. Indeed, we may assume that, in the escape depth of SE, the ions trajectories are straight lines and their energy loss is negligible. With these two assumptions, it is easy to incorporate the angle of incidence, $\theta_i = \arccos \mu_i$, into the model because the independent source can be written:

$$Q(E, \mu) = \sum_{l=0}^{\infty} \frac{2l+1}{2} \frac{P_l(\mu_i)}{|\mu_i|} Q_l(E) P_l(\mu) . \tag{4.10}$$

Due to the linearity of the Boltzmann equation, the exact solution for any angle of incidence is a linear superposition of the solutions $\psi^{(l)}(z, E, \mu)$ corresponding to a source $Q_l(E) P_l(\mu)$:

$$\phi(z, E, \mu) = \sum_{l=0}^{\infty} \frac{2l+1}{2} \frac{P_l(\mu_i)}{|\mu_i|} \psi^{(l)}(z, E, \mu) . \tag{4.11}$$

For low energy electrons, we know that the electron flux is nearly isotropic (RB). Hence, the SE yield can be approximated by:

$$\gamma(\mu_i) = \frac{\gamma_0}{|\mu_i|} + \gamma_1 \frac{\mu_i}{|\mu_i|} + \gamma_2 \frac{P_2(\mu_i)}{|\mu_i|} \tag{4.12}$$

showing that careful measurements of the yield $\gamma(\mu_i)$ as a function of the angle of incidence can provide information on the anisotropy of the source. For forward electron emission $\mu_i > 0$ while for backward emission $\mu_i < 0$.

4.2 The Age-Diffusion Model

In this model (Devooght, Dubus and Dehaes 1987a; Dubus, Devooght and Dehaes 1987), the Boltzmann equation (4.1) is transformed into a partial differential equation by replacing the exact scattering kernel by a synthetic scattering kernel which conserves the most important features of the original cross sections, the resulting Boltzmann equation being solved in the frame of the P_1 approximation

(Case and Zweifel 1966, Ferziger and Zweifel 1966) for a general space, energy and time dependent source.

The particular choice of the synthetic kernel is based upon the following considerations:

1. The elastic cross section is known to be nearly isotropic (Ganachaud and Cailler 1979a), hence it may be approximated by

$$\Sigma_{el}(E' \to E, \mu' \to \mu) \approx \frac{1}{2}\Sigma_{el}(E')\delta(E - E') \tag{4.13}$$

where $\Sigma_{el}(E')$ is the total elastic cross section.

2. The slowing down and multiplication of the electrons are due to the inelastic collisions. A rough approximation of the inelastic cross section can be obtained by assuming that the collision is spherically isotropic in the center-of-mass system, as proposed by Wolff (1954):

$$\Sigma_s(E' \to E) = \frac{\Sigma_s(E')}{E' - E_F}H(E' - E)H(E - E_F) . \tag{4.14}$$

The factor $H(E - E_F)$ is chosen to respect the fact that the Fermi sphere is completely filled, $\Sigma_s(E')$ is the total inelastic cross section.

We consider the synthetic scattering kernel defined by the limited angular expansion similar to (4.9):

$$\Sigma_s^*(E' \to E, \mu' \to \mu) = \sum_{l=0}^{N} \frac{2l+1}{2}B_l^*(E' \to E)P_l(\mu)P_l(\mu') \tag{4.15}$$

where the coefficients B_l^* are chosen in accordance with (4.13) and (4.14):

$$B_l^*(E' \to E) = \frac{H(E' - E)H(E - E_F)}{(E' - E_F)}A_l(E') + K_l(E')\delta(E - E') . \tag{4.16}$$

The unknown functions $A_l(E')$ and $K_l(E')$ are obtained by requiring that the moments of the B_l^*:

$$M_{lm}^*(E') = \int_{E_F}^{E'}(E' - E)^m B_l^*(E' \to E)dE \tag{4.17}$$

must be equal, up to some maximum values of l and m, to the corresponding moments of the B_l appearing in the Legendre polynomials expansion (4.9) of $\hat{\Sigma}_s$.

In the coherent P_1 approximation ($l = m = 1$), we obtain a diffusion-slowing down equation for the isotropic part of the electron flux:

$$-D(E)\frac{\partial^2}{\partial z^2}\phi_0(z, E, t) + \frac{1}{v}\frac{\partial}{\partial t}\phi_0(z, E, t) + \Sigma_0(E)\phi_0(z, E, t) \tag{4.18}$$

$$= \int_E^{\infty}\frac{\Sigma_{rem}(E')}{E' - E_F}\phi_0(z, E', t)dE' + Q_0(z, E, t) - \frac{1}{\Sigma_{tr}(E)}\frac{\partial}{\partial z}Q_1(z, E, t) .$$

95

To derive (4.18) a term $(1/v)\partial^2\phi_0/\partial t^2$ has been neglected since wave propagation effects are negligible compared to diffusion effects when elastic scattering is dominating all other types of scatterings. The three quantities Σ_0, $\Sigma_{tr} = 1/3D$, and Σ_{rem} are calculated from the exact scattering kernel. In this model, they characterize completely the transport process.

Equation (4.18) can easily be rewritten in three dimensions and then solved in the age-diffusion approximation. Considering a source $Q(\boldsymbol{r}, E, t) = \delta(\boldsymbol{r} - \boldsymbol{r}_0)\delta(t - t_0)Q(E)$, we obtain the Green's function in a semi-infinite medium (Devooght, Dubus and Dehaes 1987a).

The main characteristics of the Green's function are more easily discussed when it is written for an infinite medium:

$$
\begin{aligned}
\phi_{ad,\infty}(\boldsymbol{r}, E, t) &= vQ(E)f_G(|\boldsymbol{r} - \boldsymbol{r}_0|, 2vD(E)(t - t_0)) \\
&\quad \times \exp\left(-v\Sigma_0(E)(t - t_0)\right) H(t - t_0) \\
&\quad + \int_E^\infty dE' \frac{\gamma(E')}{E' - E_F} vQ(E') \\
&\quad \times \exp\left(-v\Sigma_0(E)[t - t_0 - \hat{t}_0(E' \to E)] + q(E' \to E)\right) \\
&\quad \times f_G(|\boldsymbol{r} - \boldsymbol{r}_0|, 2(vD(E)[t - t_0 - \hat{t}_0(E' \to E)] + \hat{\tau}(E' \to E))) \\
&\quad \times H(t - t_0 - \hat{t}_0(E' \to E))
\end{aligned}
\tag{4.19}
$$

where $f_G(R, \sigma^2) = \exp(R^2/(2\sigma^2))/(2\pi\sigma^2)^{3/2}$ is a three dimensional Gaussian function. $\gamma(E)$, $\hat{t}_0(E' \to E)$ and $\hat{\tau}(E' \to E)$ are known functions (Devooght, Dubus and Dehaes 1987a).

The first term in (4.19) corresponds to the diffusion, at constant energy, of the electrons emitted by the source. It decreases exponentially with time, $\Sigma_0(E)$ appearing as an effective absorption cross section.

The second term results from a three steps process:

1. A diffusion at energy E' which, in the Fermi age theory, appears only through a first flight correction included in the slowing down time $\hat{t}_0(E' \to E)$ and in the age $\hat{\tau}(E' \to E)$.

2. A continuous slowing down characterized by the age, the multiplication being accounted for through the factor $\exp(q(E' \to E))$.

3. A diffusion at energy E characterized by the diffusion coefficient $D(E)$.

The complete Green's function, $G(\boldsymbol{r}, E, t \mid \boldsymbol{r}_0, E_0, t_0)$, in a semi-infinite medium gives the isotropic part of the electron flux corresponding to a point source $\delta(\boldsymbol{r} - \boldsymbol{r}_0)\delta(E - E_0)\delta(t - t_0)$ (Devooght, Dubus and Dehaes 1987a). For a given source $Q(\boldsymbol{r}_0, E_0, t_0)$, we get the electron flux from:

$$
\phi(\boldsymbol{r}, E, t) = \int \int \int G(\boldsymbol{r}, E, t \mid \boldsymbol{r}_0, E_0, t_0) Q(\boldsymbol{r}_0, E_0, t_0) d\boldsymbol{r}_0 dE_0 dt_0
\tag{4.20}
$$

Because of the particular form of the Green's function (it is a function of $\varrho^2 = [(x - x_0)^2 + (y - y_0)^2]$ and $(t - t_0)$, we can easily obtain the Green's function

$G_{\text{et}}(z, E \mid z_0, E_0)$ corresponding to a source integrated over x_0, y_0 and t_0 and calculate the electron flux:

$$\phi(z, E) = \int_0^\infty dz_0 \int_E^\infty dE_0 \, G_{\text{et}}(z, E \mid z_0, E_0) Q(z_0, E_0) . \qquad (4.21)$$

Once we have the electron flux at the surface $\phi(0, E)$, the outgoing electron current is easily calculated in the diffusion approximation:

$$
\begin{aligned}
J(E_v, \mu_v) \;=\; & \frac{1}{2} \frac{E_v}{E_v + U_0} \phi(0, E_v + U_0) \\
& \times \mu_v \left\{ 1 + \frac{3}{2} \left[\frac{E_v \mu_v^2 + U_0}{E_v + U_0} \right]^{1/2} \frac{1 - \mu_c^2}{1 + \mu_c^3} \right\}
\end{aligned}
\qquad (4.22)
$$

where $E_v = E - U_0$ is the energy of the electron in the vacuum and $\theta_v = \arccos \mu_v$ is the angle of emission of the electron.

Obviously, the angular distribution does not follow the $J \propto \cos \theta_v$ law exactly. However, the correction is negligible if $U_0 \gg E_v$, in agreement with experimental evidence.

4.3 Integral Equations

The conversion of the Boltzmann equation into an integral equation is easy in plane geometry even for an anisotropic source. For instance the monoenergetic equation

$$\left(\mu \frac{\partial}{\partial z} + \Sigma_s \right) \phi(z, \mu) = S(z, \mu) \qquad z \geq 0 \qquad (4.23)$$

has the solution

$$
\begin{aligned}
\phi(z, \mu) \;=\; & H(\mu) Q(0, \mu) \exp\left(-\frac{\Sigma_s z}{\mu} \right) \\
& + H(\mu) \int_0^z \exp\left(-\frac{\Sigma_s (z - z')}{\mu} \right) S(z', \mu) \frac{dz'}{\mu} \\
& + H(-\mu) \int_z^\infty \exp\left(-\frac{\Sigma_s (z' - z)}{|\mu|} \right) S(z', \mu) \frac{dz'}{|\mu|} .
\end{aligned}
\qquad (4.24)
$$

The first term corresponds to an unknown boundary ($z = 0$) source chosen to satisfy the boundary condition:

$$
Q(0, \mu) = \begin{cases} \int_0^\infty \exp\left(-\Sigma_s z'/\mu \right) S(z', -\mu) dz'/\mu & \text{for } 0 \leq \mu \leq \mu_c \\ 0 & \text{for } \mu_c \leq \mu \leq 1 \end{cases} \qquad (4.25)
$$

Therefore

$$\begin{aligned}
\phi(z,\mu) &= H(\mu)H(\mu_c - \mu) \int_0^\infty \exp\left(-\frac{\Sigma_s(z+z')}{\mu}\right) S(z',-\mu)\frac{dz'}{\mu} \\
&\quad + H(\mu) \int_0^z \exp\left(-\frac{\Sigma_s(z-z')}{\mu}\right) S(z',\mu)\frac{dz'}{\mu} \\
&\quad + H(-\mu) \int_z^\infty \exp\left(-\frac{\Sigma_s(z'-z)}{|\mu|}\right) S(z',\mu)\frac{dz'}{|\mu|} \ .
\end{aligned} \tag{4.26}$$

If the source is isotropic $S(z,\mu) = S(z)$, the scalar flux $\phi(z)$ is

$$\begin{aligned}
\phi(z) &= \frac{1}{2} \int_{-1}^1 \phi(z,\mu) d\mu \tag{4.27} \\
&= \frac{1}{2} \int_0^\infty E_1\left(\frac{\Sigma_s(z+z')}{\mu_c}\right) S(z') dz' + \frac{1}{2} \int_0^\infty E_1(\Sigma_s|z-z'|) S(z') dz'
\end{aligned}$$

where E_1 is the exponential integral. If the reflection of electrons is total, $\mu_c = 1$ and

$$\phi(z) = \frac{1}{2} \int_0^\infty \left[E_1(\Sigma_s(z+z')) + E_1(\Sigma_s|z-z'|) \right] S(z') dz' \tag{4.28}$$

which amounts to write that total reflection is equivalent to an image source in $-z'$.

4.3.1 Solution of the Integral Equation in a Semi-Infinite Medium

Now if we add the energy dependence, we have

$$\begin{aligned}
\phi(z,E,\mu) &= H(\mu)H(\mu_c(E) - \mu) \\
&\quad \times \int_0^\infty \exp\left(-\frac{\Sigma_s(E)(z+z')}{\mu}\right) S(z',E,-\mu)\frac{dz'}{\mu} \\
&\quad + H(\mu) \int_0^z \exp\left(-\frac{\Sigma_s(E)(z-z')}{\mu}\right) S(z',E,\mu)\frac{dz'}{\mu} \\
&\quad + H(-\mu) \int_z^\infty \exp\left(-\frac{\Sigma_s(E)(z'-z)}{|\mu|}\right) S(z',E,\mu)\frac{dz'}{|\mu|} \tag{4.29}
\end{aligned}$$

with

$$S(z,E,\mu) = \int_E^\infty \int_{-1}^1 \hat{\Sigma}_s(E' \to E, \mu' \to \mu)\phi(z,E',\mu')dE'd\mu' + Q(z,E,\mu) \tag{4.30}$$

which is the sum of a scattering source and an independent source Q that we will assume to be spatially uniform for $z \geq 0$.

We introduce the angular expansion of ϕ and S in (4.29) and we obtain:

$$\phi_l(z,E) = \sum_{n=0}^\infty \frac{2n+1}{2} \int_0^\infty S_n(z',E)dz'$$

$$\times \left\{ (-1)^n \int_0^{\mu_c} \frac{P_l(\mu)P_n(\mu)}{\mu} \exp\left(-\frac{\Sigma_s(E)(z+z')}{\mu} \right) d\mu \right.$$
$$+ \int_0^1 \frac{P_l(\mu)P_n(\mu)}{\mu} \left(H(z-z') + (-1)^{n+l} H(z'-z) \right)$$
$$\left. \times \exp\left(-\frac{\Sigma_s(E)(|z-z'|)}{\mu} \right) d\mu \right\} . \tag{4.31}$$

From (4.30)

$$S_n(z,E) = \int_E^{\infty} B_n(E' \to E)\phi_n(z,E')dE' + Q_n(E) . \tag{4.32}$$

If we substitute (4.32) into (4.31), we obtain a system of integral equations. However we are essentially interested by $\phi_l(0,E)$ and we know also that the infinite medium solution is a good approximation (Devooght, Dubus and Dehaes 1987b). The last is solution of

$$\Sigma_s(E)\phi_\infty(E,\mu)$$
$$= \int_E^{\infty} dE' \int_{-1}^{1} \hat{\Sigma}_s(E' \to E, \mu' \to \mu)\phi_\infty(E',\mu')d\mu' + Q(E,\mu) . \tag{4.33}$$

The electron flux may be decomposed in the following way:

$$\phi(z,E,\mu) = \phi_\infty(E,\mu) + \psi(z,E,\mu) \qquad \text{for } z \geq 0 \tag{4.34}$$

where $\psi(z,E,\mu)$ is the correction taking into account the exact boundary condition. It is easy to show that the angular moments ψ_l of the flux correction are the solution of the set of integral equations:

$$\psi_l(z,E) = F_l(z,E) + \sum_{n=0}^{N} \frac{2n+1}{2} \int_0^{\infty} dz' \int_E^{\infty} B_n(E' \to E)\psi_n(z',E')dE'$$
$$\times \int_0^1 H_{nl}(\mu,z,z',E)d\mu \tag{4.35}$$

with

$$F_l(z,E) = \sum_{n=0}^{N} \frac{2n+1}{2}\phi_{\infty,n}(E)$$
$$\times \left\{ \int_0^1 \exp\left(-\frac{\Sigma_s(E)z}{\mu} \right) P_l(\mu) \left[P_n(-\mu) - P_n(\mu) \right] d\mu \right.$$
$$\left. - \int_{\mu_c}^1 P_l(\mu)P_n(-\mu) \exp\left(-\frac{\Sigma_s(E)z}{\mu} \right) d\mu \right\} \tag{4.36}$$

and

$$H_{nl}(\mu,z,z',E) = \frac{P_l(\mu)P_n(\mu)}{\mu} \left\{ (-1)^n \exp\left(-\frac{\Sigma_s(E)(z+z')}{\mu} \right) H(\mu_c - \mu) \right.$$
$$\left. + \exp\left(-\frac{\Sigma_s(E)|z-z'|}{\mu} \right) \left[H(z-z') + (-1)^{n+l} H(z'-z) \right] \right\} . \tag{4.37}$$

N is the maximum order of anisotropy of the source or the scattering cross sections. We remark that (4.35) needs the knowledge of the moments ψ_l only up to N.

Equation (4.35) together with (4.33) and (4.36) is our final result. These equations of Volterra type have a solution different from zero only if the independent terms of (4.35) are different from zero. This will be, for instance, the case if $\mu_c < 1$ which means that the electrons are not totally reflected: a spatial transient is induced by the boundary, hence $\psi_l(z, E)$, a fact that cannot be taken into account in "space-independent" methods.

Even if $\mu_c = 1$ we have the first term of the right-hand-side of (4.36). An uniform source may be anisotropic, for instance peaked in the direction of the incident ion. The resulting $\phi_\infty(E, \mu)$ will be also anisotropic although to a lesser degree because of all collisions. However, the presence of a reflecting boundary will add a new component of reverse anisotropy, hence the transient. However, if the source is uniform but even in μ, then $\phi_\infty(E, \mu)$ will necessarily be even in μ. The presence of a totally reflecting boundary will not affect the flux because of the even character of the angular distribution and $\psi_l = 0$ for all l.

Contrary to P_N methods there is no need to truncate the angular development of ϕ and system (4.35) is exact even for $l > N$, a feature common to B_N methods (Bell and Glasstone 1970; Ferziger and Zweifel 1966).

If we consider an outgoing ion, instead of an ingoing ion, we change the sign of the odd moments of $Q(E, \mu)$ (see (4.10)) and therefore of $\phi_\infty(E, \mu)$. We have therefore a theoretical mean to assess the forward-backward anisotropy of electrons emitted by an ion traversing a thin target.

An integral equation formulation was given by Puff (1964,a,b,c) who first formulated the boundary condition (4.6). However, Puff's treatment is different from ours because he uses a much simplified isotropic elastic cross section in order to get an analytical solution. The present formulation uses the knowledge of the infinite medium slowing down problem $\phi_\infty(E, \mu)$.

To calculate the outgoing current, we can use the second iterate of the integral equations (4.35). It is calculated by replacing ψ_n by the first iterate $\psi_n^{(1)}(z, E) = F_n(z, E)$ in the right-hand side of (4.35). The outgoing electron current is defined by:

$$
\begin{aligned}
J(E) &= \int_{\mu_c}^1 \phi_\infty(E, -\mu)\mu d\mu + \int_{\mu_c}^1 \psi(0, E, -\mu)\mu d\mu \\
&= J_\infty(E) + \Delta J(E) .
\end{aligned}
\tag{4.38}
$$

From (4.29) and (4.36), it is easy to show that ΔJ is given by

$$
\begin{aligned}
\Delta J(E) &= -\int_{\mu_c(E)}^1 \mu d\mu \int_E^\infty dE' \int_0^1 \hat{\Sigma}_s(E' \to E, \mu' \to -\mu)\phi_\infty(E', \mu') \\
&\quad \times \frac{\mu'}{\mu' \Sigma_s(E) + \mu \Sigma_s(E')} d\mu' \\
&\quad + \int_{\mu_c(E)}^1 \mu d\mu \int_E^\infty dE' \int_0^{\mu_c(E')} \hat{\Sigma}_s(E' \to E, \mu' \to -\mu)\phi_\infty(E', -\mu') \\
&\quad \times \frac{\mu'}{\mu' \Sigma_s(E) + \mu \Sigma_s(E')} d\mu'
\end{aligned}
\tag{4.39}
$$

$$= -\sum_{n=0}^{N}\sum_{l=0}^{N}\frac{2n+1}{2}\frac{2l+1}{2}(-1)^n \int_{E}^{\infty} dE' B_n(E' \to E)\phi_{\infty,l}(E')$$
$$\times \left[I_{nl}(E,E',1) - (-1)^l I_{nl}(E,E',\mu_c(E'))\right] \tag{4.40}$$

with

$$I_{nl}(E,E',\mu_{\max}) = \int_0^{\mu_{\max}} d\mu' \mu' P_n(\mu') P_l(\mu')$$
$$\times \int_{\mu_c(E)}^{1} d\mu \frac{\mu P_n(\mu)}{\mu \Sigma_s(E') + \mu' \Sigma_s(E)} . \tag{4.41}$$

Equation (4.39) gives the first order transport correction to the current as a sum of two terms. Both are first flight corrections, the first term corresponds to a vacuum boundary condition and the second takes into account the partial reflection.

The integration over μ and μ' in (4.41) can easily be performed analytically, while the integration over E' in (4.40) must be done numerically since B_n and $\phi_{\infty,l}$ come from a numerical calculation.

4.3.2 Transport-Albedo Model

We examine now an alternate method of solving the stationary Boltzmann equation (4.5) by use of the surface Green's function $G_S(E,-\mu \mid E_0,\mu_0)$, i.e. the boundary flux in $E,-\mu$ (the outgoing flux) for a unit boundary source in E_0,μ_0.

The boundary condition (4.6) can be satisfied if we assume an infinite medium problem for $\psi(z,E,\mu)$ defined by (4.34), i.e.

$$\left[\mu\frac{\partial}{\partial z} + \Sigma_s(E)\right]\psi(z,E,\mu) = \int_{E}^{\infty} dE' \int_{-1}^{1} d\mu' \hat{\Sigma}_s(E' \to E, \mu' \to \mu)$$
$$\times\psi(z,E',\mu') + q(E,\mu)\delta(z) \tag{4.42}$$

where the surface source is

$$q(E,\mu) = \begin{cases} -\mu\phi_\infty(E,\mu) & \mu_c \le \mu \le 1 \\ \mu\left[\phi(0,E,-\mu) - \phi_\infty(E,\mu)\right] & 0 \le \mu \le \mu_c(E) \\ 0 & \mu < 0 \end{cases} . \tag{4.43}$$

The solution of (4.42) is

$$\psi(0,E,-\mu) = -\int_{E}^{\infty} dE_0 \int_0^1 G_S(E,-\mu \mid E_0,\mu_0)\phi_\infty(E_0,\mu_0)d\mu_0$$
$$+ \int_0^{\infty} dE_0 \int_0^{\mu_c(E_0)} G_S(E,-\mu \mid E_0,\mu_0)\phi(0,E_0,-\mu_0)d\mu_0 . \tag{4.44}$$

No closed expression exists for the Green's function.

We can give an approximate solution of (4.42) if we make the following approximation

$$\int_{E}^{\infty} dE' B_n(E' \to E)\psi_n(z,E') \approx \int_{E}^{\infty} B_n(E' \to E)\psi_n(z,E)\frac{\phi_{\infty,n}(E')}{\phi_{\infty,n}(E)}$$
$$\approx \psi_n(z,E)\left[\Sigma_s(E) - \frac{Q_n(E)}{\phi_{\infty,n}(E)}\right] \tag{4.45}$$

the last relation being obtained from (4.33).

We have now a monoenergetic problem with a scattering law given by

$$\Sigma^*(E; \mu' \to \mu) = \Sigma_s(E) \sum_{n=0}^{N} \frac{2n+1}{2} P_n(\mu) P_n(\mu') \left[1 - \frac{Q_n(E)}{\Sigma_s(E)\phi_{\infty,n}(E)} \right] \quad (4.46)$$

where E appears just as a parameter.

The corresponding surface Green's function $G_S(E; -\mu, \mu_0)$ has been given explicitly by Horak and Chandrasekhar (1961), for $N = 2$, by means of "invariant imbedding" arguments (Chandrasekhar 1960).

Using their angular convention, where the angle is measured with respect to the exterior normal, the Green's function is

$$G_S(E; -\mu, \mu_0) = \frac{1}{2\mu_0} S^{(0)}(\mu_0, \mu)$$

$$\left(\frac{1}{\mu_0} + \frac{1}{\mu} \right) S^{(0)}(\mu_0, \mu) = H^{(0)}(\mu_0) H^{(0)}(\mu) K(\mu, \mu_0) \quad (4.47)$$

where $H^{(0)}(\mu)$ is the solution of the well known integral equation

$$H^{(0)}(\mu) = 1 + \mu H^{(0)}(\mu) \int_0^1 \frac{\psi^{(0)}(\mu')}{\mu + \mu'} d\mu' . \quad (4.48)$$

$\psi^{(0)}$ is a polynomial and K a function of μ and μ_0, both functions depend upon the anisotropy of the scattering law.

4.4 Numerical Solution of the Boltzmann Equation

In this section, we will present a brief introduction to the numerical calculation of the solution of the linear Boltzmann equation. For sake of simplicity, we will only consider the stationary transport equation in plane geometry with spatially uniform cross sections (see (4.5) where the time derivative is omitted). More detailed information can be found in Lewis and Miller (1984) and O'Dell and Alcouffe (1987), who discuss the numerical methods used in nuclear reactor theory.

Among the available methods, the S_N-multigroup method is considered as the most efficient computational algorithm. This method involves discretization of the three independent variables of the transport equation: the energy E, angle variable μ and space z.

For the energy variable, the multigroup method is used to evaluate the scattering source, i.e. the integral over the energy appearing on the right-hand-side of (4.5). The energy domain is partitioned in G intervals of width ΔE_g and the original transport equation is reduced to a set of G equations:

$$\left(\mu \frac{\partial}{\partial z} + \Sigma_{s,g}(\mu) \right) \phi_g^{(k+1)}(z, \mu) \quad (4.49)$$

$$= \sum_{g'=1}^{g} \int_{-1}^{1} \hat{\Sigma}_{s,g' \to g}(\mu' \to \mu) \phi_{g'}^{(k)}(z, \mu') d\mu' + Q_g(z, \mu) \qquad g = 1, \dots, G$$

where ϕ_g and Q_g are the total flux and source, respectively, in the energy interval ΔE_g; increasing g represents decreasing energy. The multigroup cross sections $\Sigma_{s,g}$ and $\hat{\Sigma}_{s,g'\to g}$ are weighted averages of the original cross sections. If the number of groups is large enough, we can simply choose $\Sigma_{s,g}(\mu) = \Sigma_s(E_g, \mu)$ and $\hat{\Sigma}_{s,g'\to g}(\mu' \to \mu) = \hat{\Sigma}_s(E_{g'} \to E_g, \mu' \to \mu)$. However, this approximation is probably not the best choice.

It is worth noting that the scattering kernel involves only scattering from high energies to lower energies. Therefore, the flux ϕ_g in group g depends only upon the fluxes in the upper groups ($g' = 1, \ldots, g$). This suggest to solve (4.49) a group at a time, starting from $g = 1$.

The upper index k of ϕ_g indicates the iterative method of solution: from an initial guess $\phi_g^{(0)}$, the scattering source is calculated and the set of equations is solved giving $\phi_g^{(1)}$ which serves to update the scattering source. This iteration process is performed until a suitable convergence criterion is satisfied.

The angular discretization method is that based upon the method of discrete ordinates proposed by Carlson (see Bell and Glasstone 1970). This method consists of evaluating the flux, solution of (4.49), only at N discrete values of μ in the interval $-1 \le \mu \le 1$, the quadrature approximation to the integral term being compatible with the choice of the μ_n's. The scattering source is approximated by:

$$\int_{-1}^{1} \Sigma_{s,g'\to g}(\mu' \to \mu_n)\phi_{g'}(z, \mu')d\mu' \approx \sum_{i=1}^{N} w_i \hat{\Sigma}_{s,g'\to g}(\mu_i \to \mu_n)\phi_{g'}(z, \mu_i) \quad (4.50)$$

where w_i are the weights of the quadrature formula. Most authors use the Gauss-Legendre integration scheme which has the property to integrate exactly a polynomial of degree $2N - 1$ using only N points. The compatibility mentioned above means that the μ_n's and w_n's must be the Gauss-Legendre ordinates and weights, respectively.

The final step consist to discretize the spatial variable and to solve iteratively the set of N differential equations:

$$\left(\mu_n \frac{\partial}{\partial z} + \Sigma_{s,g}(\mu_n)\right)\phi_g^{(k+1)}(z, \mu_n) \quad (4.51)$$

$$= \sum_{i=1}^{N} w_i \hat{\Sigma}_{s,g\to g}(\mu_i \to \mu_n)\phi_g^{(k)}(z, \mu_i)$$

$$+ \sum_{g'=1}^{g-1}\sum_{i=1}^{N} w_i \hat{\Sigma}_{s,g'\to g}(\mu_i \to \mu_n)\phi_{g'}(z, \mu_i)$$

where we have split the scattering source into two terms. Only the first one must be updated during the iteration process because the second term involves only the $\phi_{g'}$'s, for $g' < g$, which are already known.

In principle, the equations (4.51) can be solved exactly once the right-hand-side is a known function of z and the initial conditions are given. These conditions are derived from the boundary conditions. At $z = 0$, the flux must satisfy the partial reflection boundary condition (4.6) and at a distance $z_{\max}$, far inside the solid, we may assume that $\phi_g(z_{\max}, \mu_n)$ is equal to zero, at least for $\mu_n < 0$. If the target is very thin, this latter condition must be replaced by a partial reflection boundary condition, $z_{\max}$ is then the thickness of the target. This case is not very common and it will not be considered below.

In summary, the following procedure can be used to calculate $\phi_g(z, \mu_n)$ assuming that the flux in the upper groups are already known (N is assumed to be even and $\mu_n = -\mu_{N-n+1}$).

1. The scattering source is calculated from the initial guesses $\phi_g^{(0)}(z, \mu_n)$. In most problems, these initial guesses can be chosen more or less arbitrarily because the convergence of the iteration process is rather fast.

2. The equations corresponding to $\mu_n < 0$ are solved, starting from $z = z_{\max}$ with $\phi_g^{(k+1)}(z_{\max}, \mu_n) = 0$, giving $\phi_g^{(k+1)}(z, \mu_n)$ for $0 \leq z \leq z_{\max}$.

3. Then, for $0 < \mu_n \leq 1$, the integration is performed from $z = 0$ to $z_{\max}$ with $\phi_g^{(k+1)}(0, \mu_n) = \phi_g^{(k+1)}(0, -\mu_{N-n+1})$ for $0 < \mu \leq \mu_c$ and $\phi_g^{(k+1)}(0, \mu_n) = 0$ for $\mu_c < \mu \leq 1$.

4. If the convergence criterion is not satisfied, the scattering source is updated and we proceed with the next iteration (step 2).

A method, connected to the S_N method, has been used by Bindi, Lantéri and Rostaing (see for instance Bindi, Lantéri and Rostaing 1980a) and applied to SEE. However they did not take full advantage of the S_N method.

No other attempt has been made for solving the transport equation by a purely numerical method to study the SE process. However a large number of problems in high energy electron transport have been treated by S_N methods or related methods adapted to take into account high anisotropy of the cross sections (Filippone 1988).

We think that, in case where we are only interested in the electron flux or outgoing electron current, the computational effort involved either in the integral method (4.29) or in the S_N-multigroup method is much lower than in the Monte Carlo method. In our opinion, the success of the S_N method in neutron transport justify further studies to apply it to low energy electron transport.

4.5 The Model of Schou

The theory proposed by Schou (1980,1988) is an application of the models developed in sputtering theory to electron emission induced by electrons or ions in thick or thin targets. From the multispecies Boltzmann equation and power law cross sections, Schou derived an approximate connection between the electron emission characteristics and the energy deposited into electrons in the solid.

One of the most interesting feature of this model is that most quantities appearing in the final results are available from experiments or from theoretical calculations. For instance, the SE yield δ from electron or ion bombardment is determined by the surface value of the average energy $D(E, z, \cos \theta_i)$ deposited into kinetic energy of the electrons:

$$\delta = D(E, 0, \cos \theta_i) \Lambda \tag{4.52}$$

where E is the kinetic energy of the primary particles and θ_i the angle of incidence. The parameter Λ depends only upon the material:

$$\Lambda = \frac{c}{4} \int_{U_0}^{\infty} \frac{dE_0}{E_0 |dE_0/dz|} \left(1 - \frac{U_0}{E_0} \right) \tag{4.53}$$

where the target properties appear through the magnitude of the surface barrier U_0 and the low energy stopping power of electrons $|dE_0/dz|$.

Schou has shown that the deposited electronic energy is nearly proportional to the electronic stopping power of the primary particle:

$$D(E, 0, \cos \theta_i) = \beta |dE/dz|_e \tag{4.54}$$

where β depends upon the angle of incidence and the type of incident particle but is a very slowly varying function of E.

5. Theoretical Results

In this section, we will mainly consider the electron emission from polycrystalline aluminum induced by incident electrons and protons. However, we will also discuss some results for noble metals, gold being taken as an example. For incident electrons, most of these results have been obtained by MTC simulation. For incident protons, only the results obtained by the "infinite medium slowing-down" and "transport-albedo" models will be described.

We will first discuss the electron emission from a polycrystalline aluminum target, restricting the study to a primary electron energy domain from 100 eV to 1 keV. Only normal incidence of the primary beam will be considered though quite interesting complementary information can be gained by varying the angle of incidence. Some older results for gold, obtained by Cailler and coworkers, will also be presented.

We will also emphasize the role of the transport of the primary electron which is not taken into account in many SEE models, as in the "infinite medium slowing-down model" (see RB).

At last, we will compare the results obtained from the "infinite medium slowing-down" and "transport-albedo" models for incident protons, laying stress on the transport correction at the surface and on the ratio of the forward and backward yields for thin targets.

5.1 Electron Emission from a Polycrystalline Al Target

The theoretical model for electron interactions in polycrystalline Al has already been presented in detail elsewhere (Ganachaud 1977; Ganachaud and Cailler 1979a,b). Here we will only recall the main features of this model:

– The elastic cross section has been calculated from the Smrcka potential (1970) by PWEM.

– The individual and collective excitations within the jellium have been accounted for by the dielectric theory of Lindhard (1954).

– The excitation of surface plasmons has been taken into account. However, the width of the surface zone z_v, at the vacuum side, has been set equal to zero. The width z_s, at the medium side, has been calculated according to the following formula (Feibelman 1973):

$$z_s(E) = A(E - E_s)^{1/2} \qquad (5.1)$$

where E_s is a threshold energy for the surface plasmon excitation and A a constant. For aluminum, $E_s \approx 25$ eV and $A = 0.07$ Å eV^{-1} deduced from the approximate value of $z_s = 1.4$ Å at 500 eV.

– The bulk and surface plasmons have been assumed to decay into only one electron-hole pair. The probability to transfer an energy $\hbar\omega$ to an electron of initial energy E, inside the Fermi sphere, has been assumed to be proportional to the level densities in the initial and final states $\varrho(E) \cdot \varrho(E + \hbar\omega)$. For Al, $\varrho(E)$ is assumed to vary like $E^{1/2}$.

– The inner-shell ionization has been described by the Gryzinsky formalism (1965a,b,c). We have also assumed that the excited atomic core decays only via Auger transitions.

5.1.1 SEE Spectrum

Figure 5.1 compares the shapes of the theoretical and experimental (Roptin 1975) energy distribution of the electron current $J(E)$ in the true secondary peak region for $E_p = 300$ eV. The agreement in shape between the two curves is rather satisfactory. In particular, the energy E_{max}, at the maximum, and the full width at half maximum $w_{1/2}$ of both curves are nearly the same. A similar agreement is also observed for all other primary energies. Especially, the decrease with primary energy of E_{max} and $w_{1/2}$, observed experimentally, is well reproduced.

The shoulder seen at about 10.5 eV in the experimental spectrum has been attributed to the decay of bulk plasmons (Pillon, Roptin and Cailler 1976; Everhart et al. 1976). A similar structure, resulting from our assumption that bulk plasmons decay into one electron-hole pair, is also seen in the theoretical spectrum.

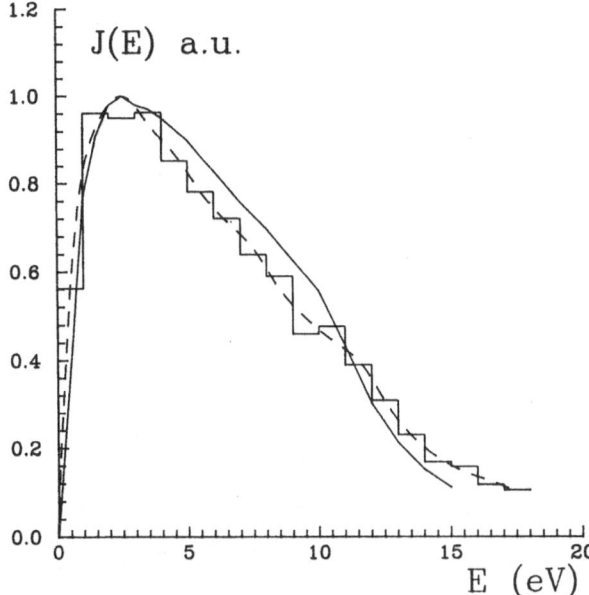

Fig. 5.1. Energy spectrum of SE for 300 eV incident electrons on Al. The histogram is the MTC result and the solid curve is the experimental spectrum obtained by Roptin (1975). The dashed curve is a smoothed MTC spectrum. The solid and dashed curves have been normalized to the same amplitude at the maximum

The dominant contributions to the SE spectrum come from the individual electron-electron collisions and from bulk plasmon decays, as seen in Fig. 5.2 and 5.3. The partial spectrum corresponding to the individual collisions determines the shape of the secondary peak. The plasmon decay contribution is rather flat up to about 12 eV, it gives rise to a broadening of the secondary peak and is responsible for the shoulder. A small difference between the positions of this shoulder in the experimental and theoretical spectrum, respectively, is clearly seen in Fig. 5.1. A better agreement can probably be obtained if more realistic plasmon dispersion curve and density of states $\varrho(E)$ are used.

The contribution of the electrons excited by the decay of a surface plasmon is much less important. The resulting structure at about 6 eV is hardly seen both in the experimental and theoretical spectra.

The contribution coming from the ionization process does not influence significantly the shape of the spectrum. However, Fig. 5.3 shows that its relative importance increases with the primary energy.

5.1.2 Energy-Loss Spectrum

In the characteristic energy-loss region, i.e. the energy domain just below the elastic peak, the electron spectrum is the energy distribution of the primary electrons which have suffered a few inelastic collisions. We compare, in Fig. 5.4, the experimental and theoretical spectra of backscattered primaries at 500 eV.

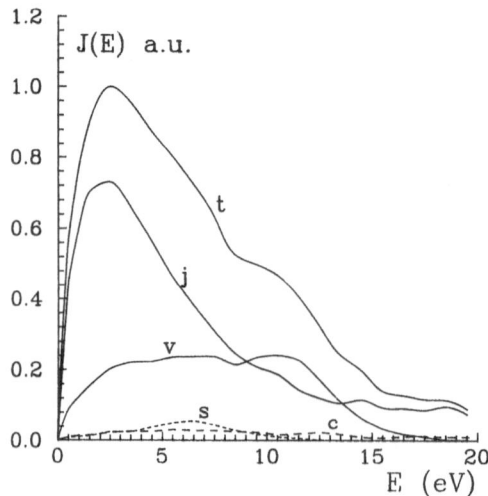

Fig. 5.2. Total energy spectrum (t) and its decomposition into partial spectra, each corresponding to a different origin of the outgoing electron: (j): individual excitation from the jellium; (v): volume plasmon decay; (s): surface plasmon decay; (c): core level excitation

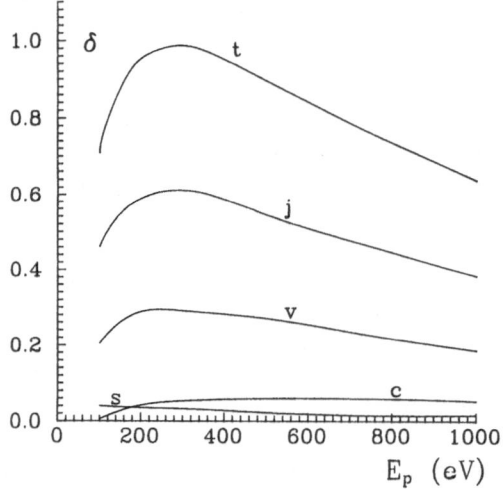

Fig. 5.3. Total SE yield (t) and partial yields (see Fig. 5.2) as a function of primary electron energy

In order to reproduce the shape of the experimental elastic peak, the incident electron spectrum has been chosen gaussian. Both spectra show similar characteristic energy loss peaks. They have been unambiguously identified (Roptin 1975) as single or multiple plasmon loss peaks. The one or two bulk plasmons loss peaks at 16.5 and 33 eV are the strongest. Weaker structures are also seen, they correspond to one (12 eV) and two (24 eV) surface plasmons excitations and to one bulk plus one surface plasmon excitations (28.5 eV).

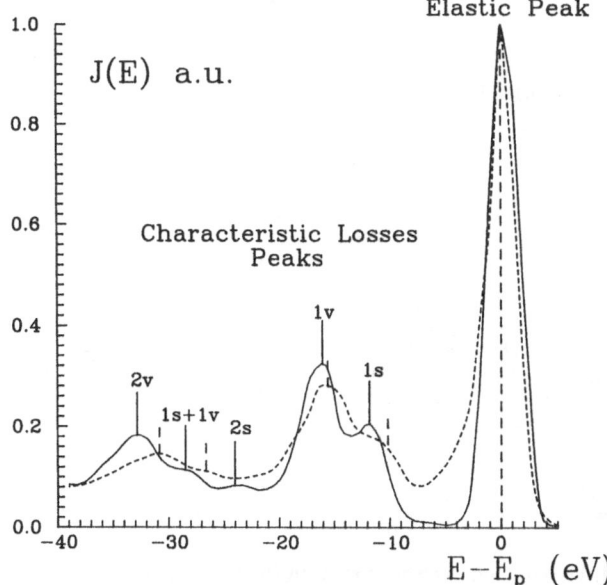

Fig. 5.4. Electron spectrum in the characteristic energy-loss region for $E_p = 500$ eV primary electrons. The solid curve is the MTC result and the dashed curve is the experimental spectrum measured by Roptin (1975). In the theoretical spectrum, the characteristic energy-loss peaks are labeled according to their origin: 1 or 2 surface (s) or bulk (v) plasmon excitations

By comparing our results to the measurements of Roptin, one can observe a good one to one correspondence between the different peaks. Nevertheless some discrepancies appear. Although the relative amplitudes of the volume and surface plasmon losses are reasonably well reproduced, the positions of the theoretical peaks are slighly shifted towards low energies. This seems to indicate that our description of the surface zone gives a good estimate of z_s but that the plasmon dispersion relations, calculated from the Lindhard theory, should be improved, as already mentioned. On the other hand, the continuous inelastic background, seen in the experimental spectrum just at the low energy side of the elastic peak, is not present in the theoretical spectrum . This shows that our model probably underestimates the importance of the individual excitations at low energy (a few eV).

We will now consider the elastic peak itself. It corresponds to backscattered primary electrons having suffered at least one elastic collision. Our theoretical estimate of the elastic coefficient ϱ_e, defined as the area of the elastic peak per incident electron, is about 0.025 at 500 eV and 0.012 at 1 keV. At 1 keV, Jablonski (1985) obtained $\varrho_e = 0.009$. He used also the MTC method, calculated the elastic cross section from a Thomas-Dirac-Fermi potential and took the inelastic mean free path from Penn (1976b) or Ashley and Tung (1982). Both estimates of ϱ_e are significantly higher than the experimental result of Schmid, Gaukler and Seiler (1983) who obtained $\varrho_e = 0.0065$. However, as pointed out by Jablonski, their experimental values were not corrected for the acceptance angle, between

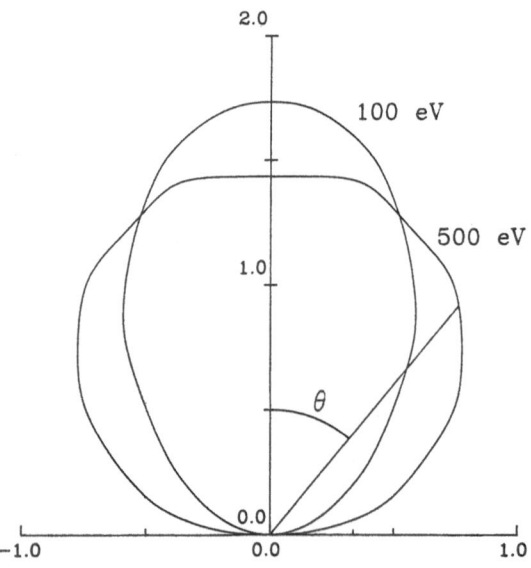

Fig. 5.5. Angular distribution of the outgoing electron current in the elastic peak region

6° and 52° instead of 0° and 90°, of the electron analyser, giving rise to an underestimation of ϱ_e.

While the angular distribution of true SE is practically not distinguishable from a cosinusoidal distribution, the angular distribution of the electrons in the elastic peak can be rather anisotropic, as seen in Fig. 5.5. For polycrystalline materials and at normal incidence, such polar diagrams are nearly ellipses. Our results show that the anisotropy is very pronounced at low primary electron energy (below about 300 eV) and tends to disappear at higher energies, in accordance with the results obtained by Jablonski (1985). This energy dependence of the anisotropy is due to the energy dependence of the mean number of elastic collisions suffered by a backscattered primary electron. At 100 eV, about 25 percent of the primary electrons suffering their first elastic collision are scattered in the backward hemisphere, whereas this percentage is only 3 percent at 1 keV. Therefore, we expect the elastic backscattering process to be mainly a single scattering process at low energy and, as already mentioned by Jablonski (1985), a multiple scattering process at higher energies.

5.1.3 Secondary Yield and Backscattering Coefficient

Figure 5.6 shows the theoretical and experimental (Bronshtein and Frajman 1969; Thomas and Pattinson 1970; Richard 1974; Roptin 1975) primary energy dependence of the true secondary yield δ and backscattering coefficient η, for normal incidence. The experimental results of Roptin (1975) were not corrected for the limited analyser acceptance angle θ_c equal to 50° with some uncertainty because the true collection geometry is probably not well known, especially at low en-

110

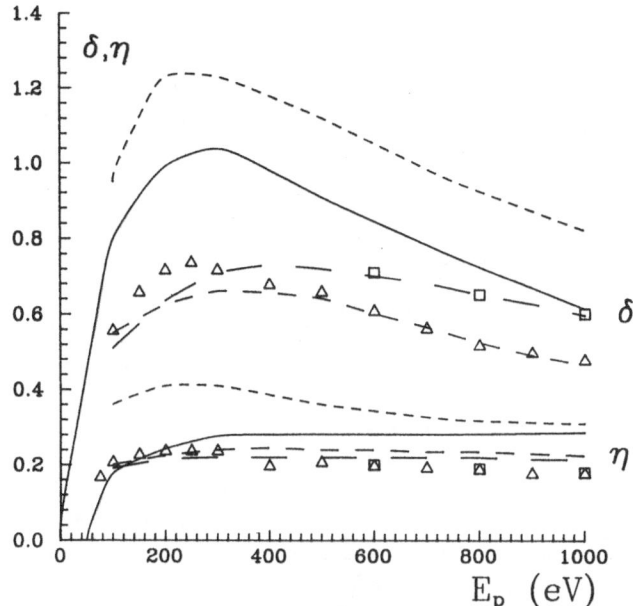

Fig. 5.6. The yields δ and η as a function of primary electron energy. Comparison between the MTC results (*solid curves*) and the experimental data: $\triangle$: Roptin (1975); *short dashed curves*: Roptin (1975) corrected for the acceptance angle; $\square$: Richard (1974); *medium dashed curves*: Bronshtein and Frajman (1969); *long dashed curves*: Thomas and Pattinson (1970)

ergy. Assuming a true cosinusoidal angular distribution, the correction is simply a factor $1/\sin^2\theta_c$. Applying this correction, we obtain the dashed curves.

Our theoretical curves lie just between the corrected and uncorrected experimental curves. This fact indicates that the precision of our theoretical estimates of δ and η is probably of the same order of magnitude as the uncertainty in the available experimental results.

In order to estimate the sensitivity of the δ and η yields to the interaction model, we have calculated these yields for various sets of elastic and inelastic cross sections. The reference yields are those calculated within the standard model described in Sect. 5.1. Figure 5.7 summarizes the results.

First we show the influence of the elastic mean free path by multiplying it by a factor two. It is clearly seen that the backscattering yield η is strongly reduced whereas the SE yield δ is only slightly modified. Increasing the elastic mfp means that the primary electron penetrates more deeply into the solid before it can be elastically backscattered. Hence the backscattering must be reduced. The SE yield is also decreased due to the deeper penetration of the primary electron but this effect is partially compensated by the increase of the total mfp at low energy.

Second we have calculated the inelastic cross sections from the dielectric functions proposed by Mermin (1970) and Vashishta and Singwi (1972) and discussed in Sect. 2.2.1. In Fig 5.7, the yields δ and η calculated using both dielectric functions are compared to those obtained with our standard model. The

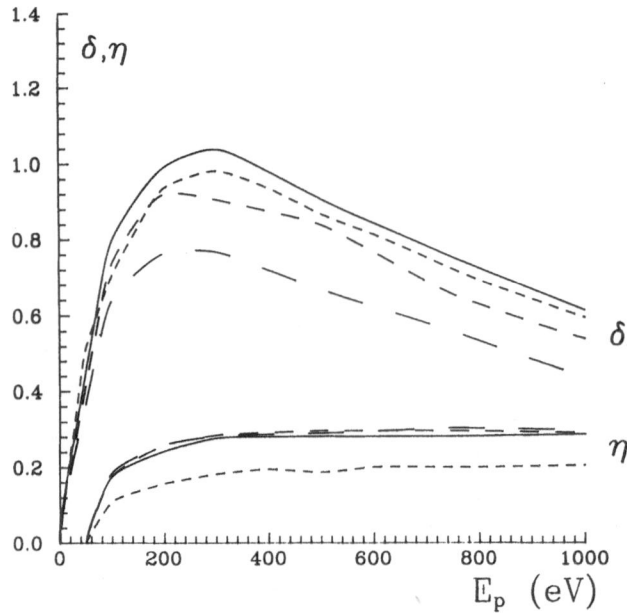

Fig. 5.7. Theoretical yields δ and η as a function of primary electron energy. The yields given by the standard model (*solid curves*) are compared to those obtained when the elastic mean free path is doubled (*short dashed curves*) and when the Lindhard dielectric function is replaced by the dielectric function given by Mermin (1970) (*medium dashed curves*) and Vashishta and Singwi (1972) (*long dashed curves*)

backscattered yield η is only slightly enhanced whereas the SE yield δ is strongly reduced. This reduction of δ is due to an increase of the inelastic mean free path in the SE energy range ($E < 50$ keV) which produces a reduction of the escape probability as a consequence.

These comparisons show that the choice of interaction cross sections has a strong influence on the calculated electron emission characteristics. This fact must be kept in mind when comparing theoretical and experimental results.

5.1.4 Spatial Distribution of Outgoing Electrons

The radial distribution of outgoing SE is the distribution of the distance of emergence of electrons with respect to the entrance point of the primary electron in the target. This distribution has been calculated for instance by Koshikawa and Shimizu (1974) and by Dubus, Devooght and Dehaes (1987).

We compare in Fig. 5.8 the radial distribution of the outgoing current $J(\varrho)$ to the partial current $J_p(\varrho)$ calculated by MTC at 300 and 600 eV. The influence of the primary electron transport appears as a broadening of the radial distribution. The mean radius of the $J(\varrho)$ distribution is 20 Å at 300 eV and 41 Å at 600 eV, whereas it is 14 Å for $J_p(\varrho)$ at both energies. The "age-diffusion" model gives also, for the mean radius of the $J_p(\varrho)$ distribution, a value of 14 Å at 300 and 600

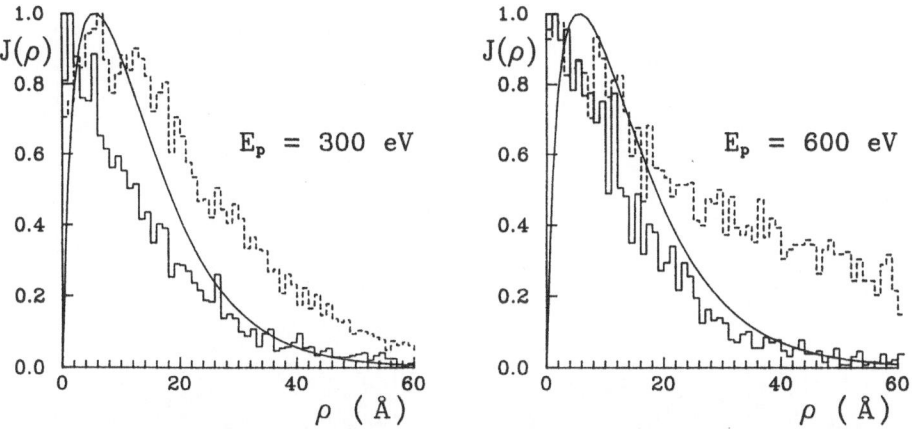

Fig. 5.8. Radial distribution of the SE current. The MTC result $J(\varrho)$ (*dashed histogram*) is shown for comparison with the current $J_p(\varrho)$ obtained by neglecting the primary electron transport: age-diffusion model (*solid curve*); MTC (*histogram*)

eV. However, the radial distribution presents a maximum at about 6 Å which does not show up in the MTC results. This discrepancy is probably due to the diffusion approximation which is not valid nearby a discontinuity.

The spatial dispersion of outgoing electrons limits the ultimate resolution of scanning electron microscopy. It has been shown by Koshikawa and Shimizu (1974) and Dubus, Devooght and Dehaes (1987) that the mean radius increases with the depth at which the electron has been excited by the primary. Asymptotically, this dependence is linear as explained by simple geometrical considerations (Dubus, Devooght and Dehaes 1987).

5.1.5 Statistical Aspects of the Electron Emission Phenomenon

We present in Fig 5.9 the distribution of the number of elastic and inelastic collisions that an electron undergoes before it escapes.

On the average, a backscattered primary electron has undergone 6.9 elastic and 4.3 inelastic collisions at 300 eV incident energy. These values increase strongly with the primary electron energy. Since the energy of the backscattered electrons is close to the incident energy, this increase in the number of collisions indicates that the mean depth of penetration of the primary increases faster than its mean free path. Indeed, the differential cross section becomes more and more forward-peaked with increasing primary electron energy.

Figures 5.9c,d show that the distributions of the number of collisions undergone by a SE are nearly energy independent. The mean number of inelastic collisions is about 0.5, whereas the mean number of elastic collisions is much larger, about 7. Again the prominent role of the elastic collisions is emphasized.

Since a SE undergoes only a few inelastic collisions, one could solve the Boltzmann equation (4.5) recursively by expanding the electron flux into partial fluxes, each corresponding to a given number of inelastic collisions. In the trans-

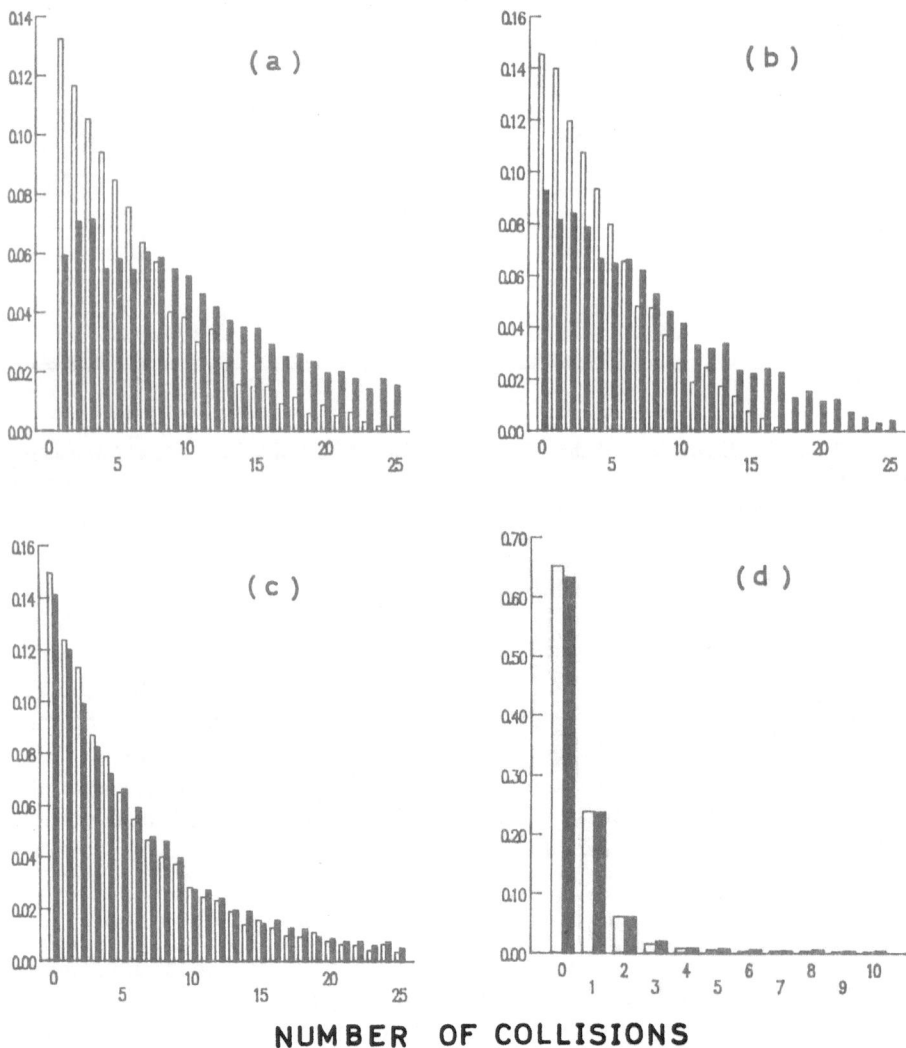

NUMBER OF COLLISIONS

Fig. 5.9. Distribution of the number of elastic and inelastic collisions that the electron experiences before it escapes. The results at 300 eV primary energy is represented as empty bars and those at 600 eV as solid bars. Backscattered electrons (contributing to η): (**a**) elastic collisions; (**b**) inelastic collisions. True secondary electrons (contributing to δ): (**c**) elastic collisions; (**d**) inelastic collisions

port equation for every partial flux, the energy appears only as a parameter and the elastic collisions are represented through a scattering kernel. Such a model is probably feasible because it incorporates the electron multiplication process and takes into account the effect of multiple elastic collisions, features not involved in the transport model proposed by Chung and Everhart (1977).

5.2 SEE from a Polycrystalline Gold Target

In this section, we present some SEE results for gold targets, a more detailed discussion can be found in Ganachaud (1977). He calculated the elastic cross section by the PWEM using the muffin-tin potential given by Ballinger and Marshall (1969). The major difference, with respect to Al, comes from the description of the inelastic collisions with the jellium, modified to include the s, p and d electrons. Ganachaud used a separable form for the energy loss function, similar to (2.15):

$$\text{Im}\left[-\frac{1}{\varepsilon(q,\omega)}\right] = \alpha_q^2 \,\text{Im}\left[-\frac{1}{\varepsilon(\omega)}\right] \tag{5.2}$$

were α_q^2 is given by (2.16). A detailed study by Cailler, Ganachaud and Bourdin (1981) confirmed that a formula similar to (5.2) allows a rather good reproduction of known values of the inelastic mfp, at least in copper. For gold, a value of $b = 0.5$ Å seems reasonable.

The function $\text{Im}[-1/\varepsilon(\omega)]$ was taken from high energy electron transmission data (Wehenkel 1975). It is worth noting that this dielectric function takes into account both the individual and collective excitations of the modified jellium. This presents no difficulties as long as bulk plasmons decay into only one electron-hole pair, as assumed in this work. The excitation of surface plasmons is completely ignored.

The inner-shell excitations, in the keV domain, were described separately in the Gryzinski formalism (1965a,b,c), including the O_1, O_2, O_3, N_{45} and N_{67} subshells (their contribution to (5.2) was approximately substracted). Furthermore Ganachaud assumed that the only relaxation mechanism is the Auger effect.

First it is important to remark that there are rather large discrepancies between the various experimental results which are available for δ and η. As observed by Pillon (1974) and Pillon et al. (1977), the measured yields and shapes of the SE spectrum change drastically during ion sputtering and annealing of the polycristalline gold target. These facts must be kept in mind when the theoretical results are compared to experimental data.

In Fig. 5.10 we compare the experimental data of Pillon (1974) and Thomas and Pattinson (1970) to a set of theoretical results obtained by the MTC method. The theoretical curves correspond to different assumptions made to describe the elastic and inelastic collisions.

Using the model described above, we obtained values of the yields (solid curves in Fig. 5.10) which have the correct order of magnitude but they are obviously not in complete agreement with the experimental data. The δ values are clearly too low above about 300 eV whereas the η values are especially overestimated at low energy. From the experience that we have acquired from the study of other metals (see also Sect. 5.1.3), we know that these discrepancies can be explained by an underestimation of the inelastic effects and a overestimation of the elastic effects.

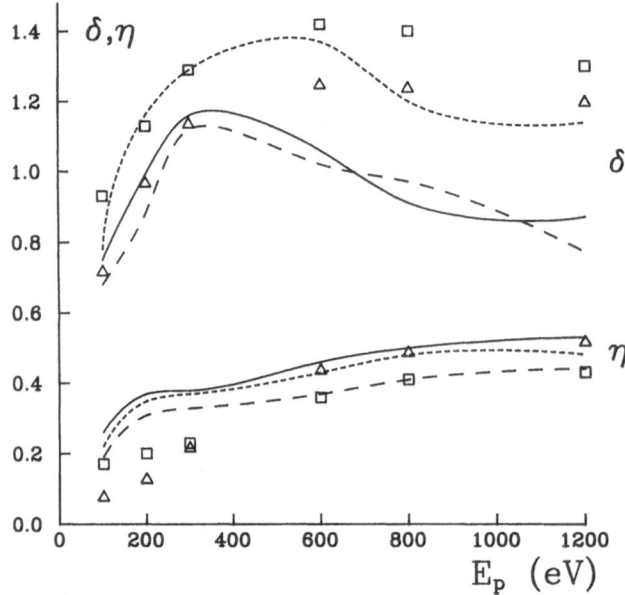

Fig. 5.10. SE yields from gold as a function of primary electron energy. Comparison between the experimental results ($\square$: Thomas and Pattinson (1970); $\triangle$: Pillon (1974)) and the MTC results: *solid curve*: basic model described in Sect. 5.2; *long dashed curve*: basic model with the elastic mean free path multiplied by 1.5; *short dashed curve*: basic model with the inelastic mean free path divided by 2

Figure 5.10 shows that the agreement between the theoretical and experimental values of η can be improved if the elastic mean free path is simply multiplied by a factor 1.5.

Because the inelastic mean free paths, calculated from the energy loss dielectric function (5.2), are in agreement with the published values, we will assume that the description of the inelastic collisions with the jellium is correct. The observed differences between the theoretical and experimental values of δ must then come from an underestimation of the ionization process.

It is known that the distribution of the ionization energy losses, estimated from the Gryzinski formula, is nearly independent of the subshell considered. The maximum always coincides with the ionization threshold. In fact, this result is quite dependent on the type of potential (coulombic, hydrogenic) used in the Gryzinski theory. More elaborate models would lead to rather different conclusions. This is in agreement with the work of Cooper (1962), Manson and Cooper (1968), Fano and Cooper (1969), for instance, as shown by Combet-Farnoux (1969) in a study of heavy atom photoionization.

Differences with the Gryzinski theory are likely to occur for subshells whose atomic quantum numbers n and l are such that $l = n - 1$. In gold, the $4f$ (N_{67}) is thus concerned. For this $4f$ subshell, the photoionization cross section exhibits a maximum at about 200 eV above the ionization threshold, in contradiction to the Gryzinski theory, and high energy transfers are thus more favoured.

Table 5.1 Variation with the energy E of the ionization mean free path l and of the energy loss ΔE for the $4f$ (N_{67}) subshell. G is for Gryzinski and P is for photoionization

E (eV)	200	400	600	800	1000
l (Å) G	71	62	71	82	94
l (Å) P	420	113	96	98	105
ΔE (eV/Å) G	1.68	2.38	2.32	2.14	1.97
ΔE (eV/Å) P	0.36	2.14	2.09	3.22	3.20

These aspects have been studied by Ganachaud (1977) who showed that the $4f$ ionization cross section deduced from photoionization data leads to a decrease of the probability of a collision but also to an increase of the mean energy loss per unit path length, at least at high energies. Table 5.1 allows a comparison between the Gryzinski and the photoionization predictions.

The consequences of these modifications have not been studied in detail. However we expect that it is mainly the stopping power which is affected by a modification of the ionization cross section. For this reason, we have just multiplied the ionization mean free path by a factor 0.5 to get a rough estimate. Figure 5.10 shows that the agreement becomes better.

We compare in Fig. 5.11 the theoretical and experimental (Pillon 1974) energy distributions for 600 eV incident electrons, they are in rather good agreement. Especially the position of the maximum and the full width at half maximum are well reproduced.

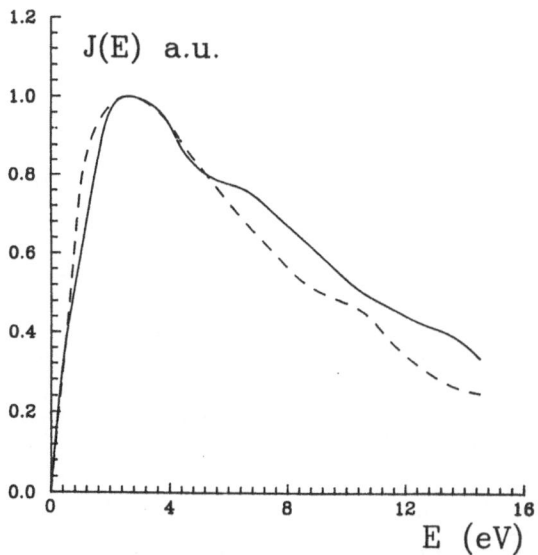

Fig. 5.11. Energy distribution of the SE from gold at 600 eV primary energy: *solid curve*: theoretical results (Ganachaud 1977); *dashed curve*: experimental data (Pillon 1974). The curves have been normalized to the same amplitude at the maximum

5.3 Influence of the Primary Electron Transport on the Secondary Electron Yield

RB have treated the electron transport problem in an approximate way. The primary electrons are assumed to follow a straight line path in the depth zone from which electrons escape. They excite electrons uniformly along their path. Hence, the yield they calculate has to be compared to the partial yield δ_p. Another approximation in their calculations consists in considering the "infinite medium slowing-down model" for the transport of secondary electrons themselves. In the following, we will compare the partial yield δ_p to the total true secondary yield δ. By using a Monte Carlo simulation code, we have calculated the yields δ and η as well as the partial yield δ_p for several values of the primary electron energy. In these calculations, the surface zone and the ionizing collisions have been neglected.

We show in Fig. 5.12 the energy dependence of the yields δ, η and δ_p for electrons incident on a polycrystalline Al target. For very low energies ($E_p \leq$ 100 eV), the partial yield δ_p is larger than the true secondary yield δ while it is smaller for higher energies. Due to the assumptions in its calculation, the partial yield δ_p is more or less proportional to the probability for electron excitation by the primary electrons. When the primary energy is below 100 eV, it excites only a few electrons and its depth of penetration into the target can be less than the depth from which SE can escape, hence $\delta \ll \delta_p$. At higher primary energies, the incident electrons penetrate deeper in the target. Some of them are backscattered and on their way back to the surface, they excite internal

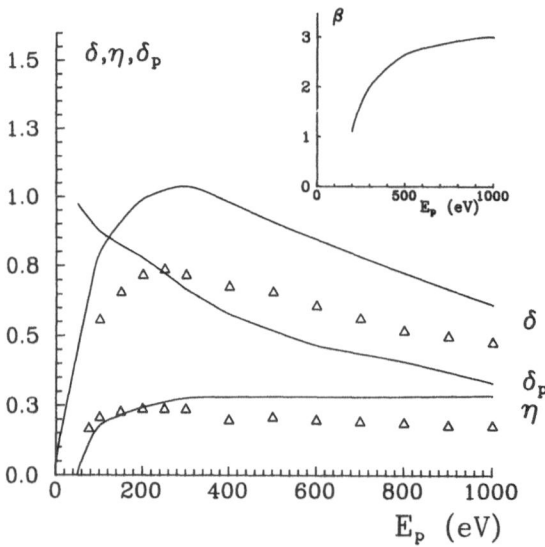

Fig. 5.12. Calculated primary energy dependence of the yields δ, δ_p and η from Al compared to the experimental values ($\triangle$) of the yields δ and η measured by Roptin (1975). The inserted figure show how the efficiency parameter β, calculated from (5.3), varies with energy

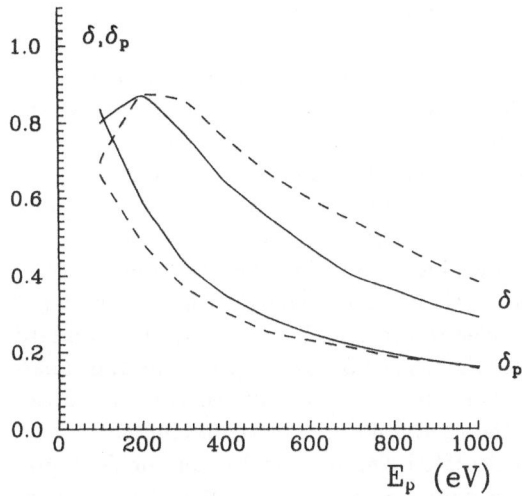

Fig. 5.13. Comparison between the yields δ and δ_p from Al calculated by the age-diffusion model (*solid curves*) and by the MTC model (*dashed curves*)

secondaries with a larger efficiency than the incident primaries. Dobretsov and Matskevitch (1957) have introduced the following well-known formula:

$$\delta = \delta_p(1 + \beta\eta) \qquad (5.3)$$

where β is an efficiency factor for electron excitation by the backscattered primaries with respect to the incident ones. The value of β deduced from the MTC calculations is about 3 for 1 keV incident electrons. This value has to be compared to the experimental results of Thomas and Pattinson (1970) who obtain values around 6 at 1 keV and to the theoretical results of Bindi et al. (1980) who obtain a value of about 5. We can also compare our δ_p value to the value obtained by RB. We obtain $\delta_p = 0.33$ for 1 keV incident electrons while RB obtain $\delta_p = 0.2$.

We have also calculated the electron yields δ and δ_p with the "age-diffusion" model (see Fig. 5.13). The internal electron source used for the calculation of δ has been obtained by Monte Carlo (ionizing collisions have been neglected here). The agreement between the Monte Carlo and the "age-diffusion" results is rather good.

It is clear from these results that the primary electron transport and backscattering must be taken into account in order to reproduce correctly the electron emission yields. This can be done by Monte Carlo or by using, for instance, a Spencer-Lewis model to study primary electron transport.

5.4 Proton Induced Electron Emission

In this section, we will present some preliminary results for proton induced electron emission from polycrystalline aluminum targets. We have neglected the charge exchange processes and assumed that the proton keeps its energy and direction unchanged in the depth from which SE can escape. Furthermore, we will consider only the inelastic collisions of the incident protons with the jellium. These collisions are described in the frame of the Lindhard dielectric function (1954), assuming that plasmons decay via interband transitions into one electron. The individual proton-electron cross sections are calculated using the model of Brice and Sigmund (1980). For the electron interactions, we use the standard model described in Sect. 5.1 except for the neglect of surface plasmons and inner-shell collisions. In spite of these rather crude approximations, our results can provide some useful information on the proton induced electron emission.

We compare in Fig. 5.14 the calculated electron yield γ for incident protons on polycrystalline Al targets to experimental results (Baragiola, Alonso and Oliva-Florio 1979; Svensson, Holmen and Buren 1981; Hasselkamp et al. 1981). The yield has been calculated with the "age-diffusion", "infinite medium slowing-down" and "transport-albedo" ($l = 2$) models. There is a general agreement between the theory and the experiments. The value of the proton energy for which the yield is maximum is correctly predicted (55 keV). For higher ener-

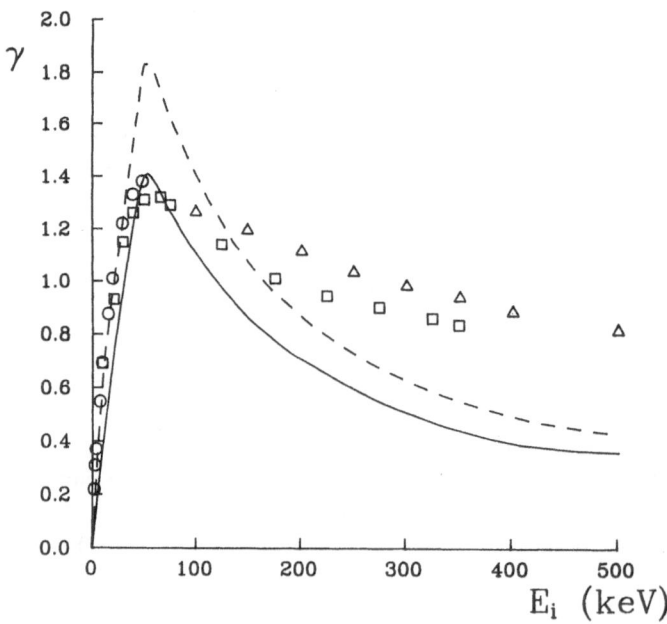

Fig 5.14. Electron yields γ induced by protons incident on a polycrystalline Al target as a function of proton energy: *solid curve*: transport-albedo model; *dashed curve*: infinite medium slowing down model; $\bigcirc$ (exp): Baragiola, Alonso and Oliva Florio (1979); $\square$ (exp): Svensson, Holmen and Buren (1981); $\triangle$ (exp): Hasselkamp et al. (1981)

gies ($E \geq 200$ keV), the calculated yield is smaller than the experimental results. This disagreement can be due to the neglect of inner-shell ionizations by the incident protons. The surface correction in the "transport-albedo" model, i.e. the reduction of the calculated yield compared to the "infinite medium slowing-down" model is about 20%, a value that is in good agreement with preliminary Monte Carlo calculations (these calculations predict a surface correction of about 25%).

The absolute electron energy spectra calculated with the "infinite medium slowing-down" and "transport-albedo" ($l = 2$) models are compared in Fig 5.15 to the absolute experimental data of Hasselkamp and Scharmann (1983). The small disagreement between theoretical and experimental results can be due once again to the neglect of inner-shell ionizations by the incident protons. The energy dependent surface correction increases with the outgoing electron energy. There is almost no correction at very low energies at which the potential barrier reflects almost all electrons. For higher energies, the potential barrier becomes transparent and the boundary condition becomes a vacuum boundary condition.

These results show that the surface correction, i.e. the correction due to the decrease of the internal electron flux near the vacuum-medium interface is not negligible and that the "infinite medium slowing-down" model overestimates the outgoing electron yield γ by about 25%.

In thin foil experiments, the protons are able to pass through the foil and SE are emitted from the forward and backward surfaces. The forward electron yield γ_F is larger than the backward electron yield γ_B. We compared in Fig. 5.16 the forward-backward yield ratio $R_\gamma = \gamma_F/\gamma_B$ calculated with the "infinite medium slowing-down" and the "transport-albedo" models to the experimental data of

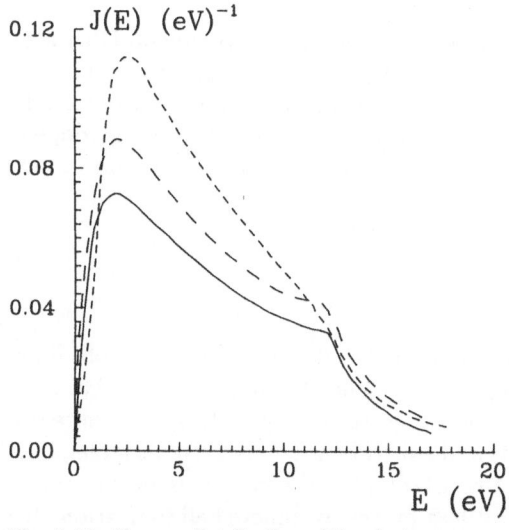

Fig. 5.15. Energy distribution of the electron current for 200 keV protons incident on Al: *solid curve*: transport-albedo model; *long dashed curve*: infinite medium slowing down model; *short dashed curve*: experimental data (Hasselkamp and Scharman 1983)

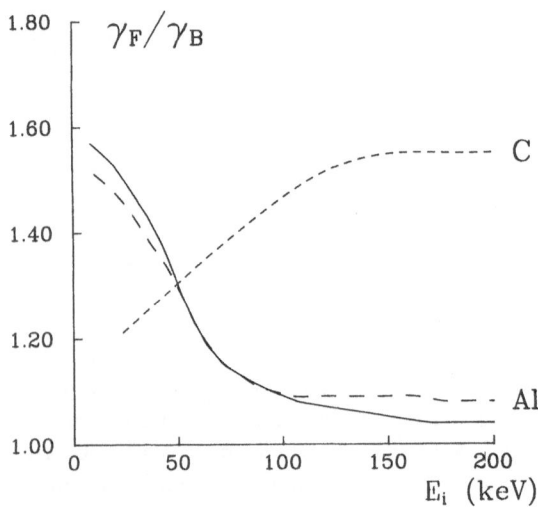

Fig 5.16. Forward-backward yield ratio as a function of proton energy: *short dashed curve*: experimental data for a thin carbon target (Meckbach, Braunstein and Arista 1975); *long dashed curve*: infinite medium slowing down model for Al; *solid curve*: transport-albedo model for Al

Meckbach, Braustein and Arista (1975) for protons incident on amorphous carbon targets. The internal electron source used for the calculations is the same for both the forward and backward emissions. Though the materials are not the same, we expect the general trends to be similar for Al and carbon targets. Preliminary calculations for electron emission from amorphous carbon targets using a free-electron model for carbon give indeed results in good agreement with experiments (Dubus 1987).

The obvious disagreement between the theoretical and experimental forward-backward yield ratios remains to be explained. However, preliminary calculations show that the electron capture and loss processes in an Al target decrease strongly this ratio for proton energies lower than 100 keV. As a result, the energy dependence of this ratio becomes similar to the experimental result for a carbon target. Further work should be done to confirm these results.

5.5 Conclusions

We have given in this chapter an overview of the results which have been obtained for incident electrons and protons on polycrystalline aluminum targets. We have also discussed some older results for incident electrons on gold targets. There is a general agreement between our results and experiments. Observed disagreements are probably due to the neglect of important physical processes in our set of interaction cross sections (capture and loss processes, inner-shell ionizations for incident protons, ...). Work is in progress to incorporate these processes in our calculations.

6. Conclusions and Future Prospects

We have given in the present work an overview both on the electron-solid interaction models that were used in SEE calculations and on the low energy electron transport models. The transport models are the Monte Carlo simulation methods and the numerical or semi-analytical solutions of the Boltzmann equation.

We have discussed the results that we have obtained for electron (SEE) and proton (IIEE) induced secondary emission from aluminum. We have also given some results for SEE from gold targets.

There is a general agreement between these results and the experimental data, although some discrepancies still remain. In particular, the ratio of the forward and backward yields for protons passing through thin targets is in disagreement with the existing experimental data. Preliminary calculations have shown that a better agreement can be obtained if the electron capture and loss processes are included in the model.

We have also compared the results obtained using several electron interaction models in aluminum and emphasized the important role played by the elastic collisions.

It is worth noting that the results given by our approximate models are in good agreement with those obtained by Monte Carlo simulations, considered to be "exact". Up to now, all authors have developed their model for SEE and IIEE by choosing both a set of cross sections and a transport model. But no systematic comparisons of the transport models have been made with the same set of cross sections. In particular, it should be interesting to carefully compare the approximate models described in the present work to the S_N and MTC models which can be considered as exact.

Many important applications of the secondary electron emission from solids can be found, as for instance: scanning electron microscopy, Auger electron spectroscopy and electron multipliers. Particular interest should be paid to the detection of fine-structures superimposed on the true secondary peak which can provide useful information about the transitions in the electronic structure of solids. Cailler and Ganachaud (1990a) have recently reviewed the different mechanisms which have been proposed for the creation of these fine structures. They quoted diffraction phenomena, bulk and surface plasmon decay, interband transitions to unoccupied levels, Auger transitions, Fano autoionization emission and surface wave-matching effects. As reported above, some of them were satisfactorily taken into account in a quantitative description of secondary electron emissive properties. This is especially true for the contributions of bulk and surface plasmons. We can also state that the experimentally determined optical loss function accounts for some of the effects connected to the unoccupied final state density. In contrast, contributions from the Fano autoionization emission were not, up to now, taken into account. However, papers by Nygaard (1975), Cornaz et al. (1987), Aebi et al. (1987), Erbudak et al. (1987) and Palacio, Sanz and Martinez-Duart (1987) suggest that these contributions could be significative in the heavy alkali metals and the less-than-half full d-shell transition metals. In the future, it should be very interesting to reach for different kind of materials a state of comprehension similar to that reached for NFE metals. This would also be particularly important for insulators to account for the very high yield values of these materials. However, in such a case, the problem would be much more intricate due to charge effects.

Probably the fine-structures superimposed on the true secondary peak should be much more evident in angle-resolved secondary electron distributions. Unfortunately, only a few number of experimental results of such measurements can be found in the literature (see Cailler and Ganachaud, 1990a) and there is a need for more numerous results. This will undoubtely result in further progress.

Some important progresses could also be made by using recent efficient techniques developed in Computer Science and data processing.

References

Adesida I., Shimizu R., Everhart T.E. [1978]: Appl. Phys. Lett. **33**, 849. (Sect. 2.1)

Aebi P., Erbudak M., Leonardi A., Vanini F. [1987]: J. Electron Spectrosc. Related Phenom. **42**, 351. (Chap. 6)

Akkerman A.F., Gibrekhterman A.L. [1985]: Nucl. Instr. Meth. **B6**, 496. (Chap. 3, Sects. 3.2, 3.3)

Anderson S.L. [1990]: SIAM Review **32**, 221. (Sect. 3.4)

Ashley J.C., Tung C.J. [1982]: Surf. Interf. Anal. **4**, 52. (Sect. 5.1.2)

Ballinger R.A., Marshall C.A.W. [1969]: J. Phys. **C2**, 1822. (Sect. 5.2)

Baragiola R.A., Alonso E.V., Oliva-Florio A. [1979]: Phys. Rev. **B19**, 121. (Sect. 5.4)

Baroody E.M. [1950]: Phys. Rev. **78**, 780. (Chap. 1)

Baroody E.M. [1956]: Phys. Rev. **101**, 1679. (Sect. 2.3.2)

Beaglehole D. [1965]: Proc. Phys. Soc. **85**, 1007. (Sect. 2.3.2)

Bell G.I., Glasstone S. [1970]: *Nuclear Reactor Theory* (Van Nostrand Reinhold Company, New York). (Sects. 4.3.1, 4.4)

Bellman R.E., Kalaba R.E., Prestrud M.C. [1963]: *Invariant Imbedding and Radiative Transfer in Slabs of Finite Thickness* (American Elsevier Publishing Company, Inc., New York). (Chap. 4)

Bennett A.J., Roth L.M. [1972]: Phys. Rev. **B5**, 4309. (Chap. 1)

Berger M.J. [1963]: "Monte Carlo calculation of the penetration and diffusion of fast charged particles", in *Methods in Computational Physics, Volume 1*, ed. by Alder B., Fernbach S., Rotenberg M. (Academic Press, New York) pp. 135–215. (Chap. 3)

Bethe H.A. [1941]: Phys. Rev. **59**, 940. (Chap. 1)

Bethe H.A., Rose M.E., Smith L.P. [1938]: Proc. Am. Phil. Soc. **78**, 573. (Chap. 1)

Bindi R., Lantéri H., Rostaing P. [1980a]: J. Phys. **D13**, 267. (Chap. 1, Sects. 2.1, 4.4)

Bindi R., Lantéri H., Rostaing P. [1980b]: J. Phys. **D13**, 461. (Chap. 1)

Bindi R., Lantéri H., Rostaing P. [1987]: Scanning Microscopy **1**, 1475. (Chap. 1)

Bindi R., Lantéri H., Rostaing P., Keller P. [1980]: J. Phys. **D13**, 2351. (Sect. 5.3)

Bonham R.A., Strand T.G. [1963]: J. Chem. Phys. **39**, 2200. (Sect. 2.1)

Brauer W., Rösler M. [1985]: Phys. Stat. Sol. **(b) 131**, 177. (Chap. 1)

Brice D.K., Sigmund P. [1980]: Mat. Fys. Medd. Dan. Vid. Selsk. **40:8**, 1. (Sect. 5.4)

Bronshtein I.M., Frajman B.S. [1969]: *Secondary Electron Emission (in Russian)* (Ed. Naouka, Moscow). (Sect. 5.1.3)

Brosens F., Lemmens L.F., Devreese J.T. [1976]: Phys. Stat. Sol. (b) **74**, 45. (Sect. 2.2.1)

Brown F.B., Martin W.R. [1985]: Progress in Nuclear Energy **14**, 269. (Sect. 3.5)

Bruining H. [1954]: *Physics and Applications of Secondary Electron Emission* (Pergamon Press Ltd, London). (Chap. 1)

Cailler M. [1969]: "Contribution à l'étude théorique de l'émission électronique secondaire induite par bombardement électronique"; Thèse d'Etat de l'Université de Nantes. (Chap. 1, Sect. 2.3.2)

Cailler M., Ganachaud J.P. [1972]: J. Physique **33**, 903. (Chap. 1, Sect. 2.3.2, Chap. 3)

Cailler M., Ganachaud J.P. [1990a]: to be published in Scanning Microscopy, Supplement 4. (Sect. 2.1, Chap. 3)

Cailler M., Ganachaud J.P. [1990b]: to be published in Scanning Microscopy, Supplement 4. (Chap. 6)

Cailler M., Ganachaud J.P., Bourdin J.P. [1981]: Thin Solid Films **75**, 181. (Sects. 2.3.2, 5.2)

Callaway J. [1959]: Phys. Rev. **116**, 1368. (Sect. 2.3.1)

Calogero F. [1967]: *Variable Phase Approach to Potential Scattering* (Academic Press). (Sect. 2.1)

Case K.M., de Hoffmann F., Placzek G. [1953]: *Introduction to the Theory of Neutron Diffusion, Volume 1* (Los Alamos Scientific Laboratory Report, U.S.A.E.C., Los Alamos, New Mexico). (Chap. 4)

Case K.M., Zweifel P.F. [1967]: *Linear Transport Theory* (Addison-Wesley, Reading, Massachussets). (Chap. 4, Sect. 4.2)

Cashwell E.D., Everett C.J. [1959]: *A practical manual on the Monte Carlo method for random walk problems* (Pergamon Press Ltd, London). (Sect. 3.4)

Chandrasekhar S. [1960]: *Radiative Transfer* (Dover Publications, Inc., New York). (Chap. 4, Sect. 4.3.2)

Chung M.S., Everhart T.E. [1977]: Phys. Rev. **B15**, 4699. (Chap. 1, Sects. 2.2.2, 5.1.5)

Combet-Farnoux F. [1969]: J. Physique **30**, 521. (Sect. 5.2)

Cooper J.W. [1962]: Phys. Rev. **128**, 681. (Sect. 5.2)

Cornaz A., Erbudak M., Aebi P., Stucki F., Vanini F. [1987]: Phys. Rev. **B35**, 3062. (Chap. 6)

Dehaes J.C., Carmeliet J., Berry H.G. [1989]: Phys. Rev. **A40**, 5583. (Chap. 4)

Dejardin-Horgues C., Ganachaud J.P., Cailler M. [1976]: J. Phys. C9, L633. (Sect. 2.3.2)

Devooght J., Dehaes J.C., Dubus A., Hollasky N. [1984]: "A time dependent secondary electron transport model", in *Forward Electron Ejection in Solids, Proceedings, Aarhus, Denmark 1984*, ed. by Groeneveld K.O., Meckbach W., Sellin I.A. (Springer-Verlag, Berlin) pp. 52–61. (Chap. 1)

Devooght J., Dubus A., Dehaes J.C. [1987a]: "Semianalytical theory of ion induced secondary electron emission and transport in semiinfinite media", in *International Topical Meeting on Advances in Reactor Physics, Mathematics and Computation*, pp. 339–349. (Sect. 4.3.1)

Devooght J., Dubus A., Dehaes J.C. [1987b]: Phys. Rev. **B36**, 5093. (Chap. 4, Sect. 4.2)

Ding Z.J., Shimizu R. [1988]: Surf. Sci. **197**, 539. (Chap. 1, Sects. 2.3.2, 2.3.3, Chap. 3, Sect. 3.2)

Ding Z.J., Shimizu R. [1989]: Surf. Sci. **222**, 313. (Sects. 2.2, 2.3.2)

Dubus A. [1987]: "Application de l'équation de Boltzmann au transport d'électrons de basse énergie dans les solides et notamment à l'émission d'électrons secondaires"; Thèse de Docteur en Sciences Appliquées de l'Université Libre de Bruxelles. (Sects. 3.4, 5.4)

Dubus A., Devooght J., Dehaes J.C. [1986]: Nucl. Instr. Meth. **B13**, 623. (Chap. 1)

Dubus A., Devooght J., Dehaes J.C. [1987]: Phys. Rev. **B36**, 5110. (Sects. 4.2, 5.1.4)

Dubus A., Devooght J., Dehaes J.C. [1990]: Scanning Microscopy 4, 1. (Sects. 3.5, 4.1)

Duderstadt J.J., Martin W.R. [1979]: *Transport Theory* (John Wiley and Sons Inc., New York). (Chap. 4)

Emerson L.C., Birkhoff R.D., Anderson V.E., Ritchie R.H. [1973]: Phys. Rev. **B5**, 1798. (Sect. 2.3.1)

Erbudak M., Vanini F., Sulmoni D., Aebi P. [1987]: Surf. Sci. **189/190**, 771. (Chap. 6)

Everhart T.E., Saeki N., Shimizu R., Koshikawa T. [1976]: J. Appl. Phys. **47**, 2941. (Sects. 2.2.2, 5.1.1)

Fano U., Cooper J.W. [1969]: Rev. Mod. Phys. **40**, 441. (Sect. 5.2)

Feibelman P.J. [1973]: Surf. Sci. **36**, 558. (Sects. 2.2.2, 5.1)

Feibelman P.J. [1974]: Phys Rev. **B9**, 5077. (Sect. 2.2.2)

Ferziger J.H., Zweifel P.F. [1966]: *The Theory of Neutron Slowing-down in Nuclear Reactors* (Pergamon Press Ltd, Oxford). (Sects. 4.2, 4.3.1)

Filippone W.L. [1988]: Nucl. Sci. Eng. **99**, 232. (Sect. 4.4)

Fröhlich H. [1955]: Proc. Phys. Soc. **B68**, 657. (Sect. 2.3.2)

Fry J.L. [1969]: Phys. Rev. **179**, 892. (Sect. 2.3.1)

Ganachaud J.P. [1977]: "Contribution à l'Etude Théorique de l'Emission Electronique Secondaire des Métaux"; Thèse d'Etat de l'Université de Nantes. (Chap. 1, Sects. 2.1, 2.2.2, 2.3.2, 3.4, 5.1, 5.2)

Ganachaud J.P., Cailler M. [1973]: J. Physique **34**, 91. (Chap. 1, Sect. 2.3.2)

Ganachaud J.P., Cailler M. [1979a]: Surf. Sci. **83**, 498. (Chap. 1, Sects. 2.1, 2.2.2, Chap. 4, Sects. 4.2, 5.1)

Ganachaud J.P., Cailler M. [1979b]: Surf. Sci. **83**, 519. (Chap. 1, Sect. 2.2.2, Chaps. 3, 4, Sect. 5.1)

Gay T.J., Berry H.G. [1979]: Phys. Rev. **A19**, 952. (Chap. 4)

Greisen F.C. [1968]: Phys. Stat. Sol. **25**, 753. (Sect. 2.1)

Gryzinski M. [1965a]: Phys. Rev. **A138**, 305. (Sects. 2.2.3, 2.3.2, 2.3.3, 5.1, 5.2)

Gryzinski M. [1965b]: Phys. Rev. **A138**, 322. (Sects. 2.2.3, 2.3.2, 2.3.3, 5.1, 5.2)

Gryzinski M. [1965c]: Phys. Rev. **A138**, 336. (Sects. 2.2.3, 2.3.2, 2.3.3, 5.1, 5.2)

Hammersley J.M., Handscomb D.C. [1964]: *Monte Carlo Methods* (Methuen and Co. Ltd., London). (Sect. 3.4)

Hasegawa M. [1971]: J. Phys. Soc. Japan **31**, 649. (Sect. 2.2.2)

Hasegawa M., Watabe M. [1969]: J. Phys. Soc. Japan **27**, 1393. (Sect. 2.2.2)

Hasselkamp D. [1985]: "Die ioneninduzierte kinetische Elektronenemission von Metallen bei mittleren und grossen Projektilenergien"; Habilitationsschrift, Universität Giessen. (Chap. 1)

Hasselkamp D., Lang K.G., Scharmann A., Stiller N. [1981]: Nucl. Instr. Meth. **180**, 349. (Sect. 5.4)

Hasselkamp D., Scharmann A. [1983]: Vak. Tech. **32**, 9. (Sect. 5.4)

Horak H.G., Chandrasekhar S. [1961]: Astrophys. J. **134**, 45. (Sect. 4.3.2)

Hubbard J. [1957]: Proc. Roy. Soc. **A243**, 336. (Sect. 2.2.1)

Ichimura S., Shimizu R. [1981]: Surf. Sci. **112**, 386. (Chap. 1, Sects. 2.1, 3.2)

Jablonski A. [1985]: Surf. Sci. **151**, 166. (Sect. 5.1.2)

Jablonski A. [1989]: Surf. Interf. Anal. **14**, 659. (Sect. 2.1)

Joachain C.J. [1983]: *Quantum Collision Theory* (North-Holland Publishing Company, Amsterdam). (Sect. 2.1)

Jousset D. [1987]: "Etude théorique et expérimentale des distributions énergétiques des électrons primaires, secondaires et Auger rétrodiffusés en spectroscopie des électrons Auger. Application à la caractérisation des couches minces d'alumine sur aluminium"; Thèse, Université Paris VI. (Sect. 2.1)

Kamiya Y., Shimizu R. [1976]: Jap. J. Appl. Phys. **15**, 2067. (Sect. 3.1)

Kliewer K.L., Fuchs R. [1969]: Phys. Rev. **181**, 552. (Sect. 2.2.1)

Knuth D.E. [1981]: *The Art of Computer Programming, Volume 2, Seminumerical Algorithms* (Addison-Wesley, Reading, Massachussets). (Sect. 3.4)

Koshikawa T., Shimizu R. [1974]: J. Phys. **D7**, 1303. (Chapts. 1, 3, Sect. 5.1.4)

Kugler A.A. [1975]: J. Stat. Phys. **12**, 35. (Sect. 2.2.1)

Lantéri H., Bindi R., Rostaing P. [1981]: J. Comput. Phys. **39**, 22. (Chap. 1)

Lewis E.E., W.F. Miller Jr. [1984]: *Computational Methods of Neutron Transport* (John Wiley and Sons Inc., New York). (Chap. 4, Sect. 4.4)

Lindhard J. [1954]: Mat. Fys. Medd. Dan. Vid. Selsk. **28:8**, 1. (Chap. 2, Sects. 2.2, 2.2.1, 2.2.2, 2.3, 3.4, 5.1, 5.4)

Lux I., Pázsit I. [1981]: Radiation Effects **59**, 27. (Chap. 4)

Mahan G.D. [1990]: *Many-Particle Physics* (Plenum Press, New York). (Sect. 2.2.1)

Manson S.T., Cooper J.W. [1968]: Phys. Rev. **165**, 126. (Sect.5.2)

McGrath E.J., Irving D.C. [1974]: **RSIC-38**, Oak Ridge National Laboratory (Oak Ridge). (Sect. 3.5)

Meckbach W., Braunstein G., Arista N. [1975]: J. Phys. **B8**, L344. (Sect. 5.4)

Mermin N.D. [1970]: Phys. Rev. **B1**, 2362. (Sects. 2.2.1, 5.1.3)

Mignot H. [1974]: "Extension à l'or et à l'argent d'un modèle théorique de simulation de l'émission électronique secondaire induite par bombardement électronique"; Thèse de troisième cycle de l'Université de Nantes. (Sect. 2.3.2)

Müller J., Burgdörfer J. [1990]: Phys. Rev. **A41**, 4903. (Sect. 3.1)

Nagel S.R., Witten T.A. [1975]: Phys. Rev. **B11**, 1623. (Sect. 2.3.2)

Nigam B.P., Sundaresan M.K., Wu Ta-You [1959]: Phys. Rev. **115**, 491. (Sect. 2.1)

Noumerov B.V. [1924]: Monthly Notices Roy. Astrom. Soc. **84**, 592. (Sect. 2.1)

Nozières P., Pines D. [1958]: Phys. Rev. **111**, 442. (Sect. 2.2.1)

Nygaard K.J. [1975]: Phys. Rev. **A11**, 1475. (Chap. 6)

O'Dell R.D., Alcouffe R.E. [1987]: **LA-10983-MS**, Los Alamos National Laboratory (Los Alamos, New Mexico). (Chap. 4, Sect. 4.4)

Palacio C., Sanz J.M., Martinez-Duart J.M. [1987]: Surf. Sci. **191**, 385. (Chap. 6)

Pendry J.B. [1974]: *Low Energy Electron Diffraction* (Academic Press, London and New York). (Sect. 2.1)

Penn D.R. [1976a]: J. Electron Spectrosc. Related Phenomena **9**, 29. (Sect. 5.1.2)

Penn D.R. [1976b]: Phys. Rev. **B13**, 5248. (Sect. 2.3)

Penn D.R. [1987]: Phys. Rev. **B35**, 482. (Sects. 2.2, 2.3.2, 2.4)

Pillon J. [1974]: "Etude critique d'un spectroscope Auger pour l'émission électronique secondaire. Résultats obtenus pour un cristal de cuivre (111)"; thèse de Docteur-Ingénieur de l'Université de Nantes. (Sect. 5.2)

Pillon J., Ganachaud J.P., Roptin D., Mignot H., Dejardin-Horgues C., Cailler M. [1977]: "Secondary electron emission of metal surfaces (Al,Ag,Au)", in *Proceedings 7th Intern. Vac. Congr. & 3rd Intern. Conf. Solid Surfaces*, pp. 473–476. (Sects. 2.3.2, 5.2)

Pillon J., Roptin D., Cailler M. [1976]: Surf. Sci. **57**, 741. (Sects. 2.2.2, 5.1.1)

Pines D. [1963]: *Elementary Excitations in Solids* (W.A. Benjamin Inc., New York). (Sect. 2.2)

Powell C.J. [1974]: Surf. Sci. **44**, 29. (Sect. 2.3.2)

Puff H. [1964a]: Phys. Stat. Sol. **4**, 125. Chap. 1, (Sect. 4.3.1)

Puff H. [1964b]: Phys. Stat. Sol. **4**, 365. Chap. 1, (Sect. 4.3.1)

Puff H. [1964c]: Phys. Stat. Sol. **4**, 569. Chap. 1, (Sect. 4.3.1)

Raether H. [1988]: *Surface Plasmons on Smooth and Rough Surfaces and on Gratings*, Springer Tracts in Modern Physics, Vol. 111 (Springer-Verlag, Berlin). (Sect. 2.2.2)

Richard C. [1974]: "Contribution à l'étude de l'émission électronique secondaire par bombardement électronique, sur des cibles d'Al et Au"; Thèse de Spécialité de l'Université de Provence. (Sect. 5.1.3)

Ritchie R.H., Howie A. [1977]: Phil. Mag. **36**, 463. (Sect. 2.3.2)

Ritchie R.H., Tung C.J., Anderson V.E., Ashley J.C. [1975]: Radiation Research **64**, 181. (Sect. 2.3.1)

Roptin D. [1975]: "Etude expérimentale de l'émission électronique secondaire de l'aluminium et de l'argent"; Thèse de Docteur-Ingénieur de l'Université de Nantes. (Sects. 5.1.1, 5.1.2, 5.1.3)

Rösler M. [1987]: "Theorie der Sekundärelektronenemission und ioneninduzierten kinetischen Elektronenemission einfacher Metalle"; Habilitationsschrift, Berlin. (Chap. 1)

Rösler M., Brauer W. [1981a]: Phys. Stat. Sol. (b) **104**, 161. (Chap. 1)

Rösler M., Brauer W. [1981b]: Phys. Stat. Sol. (b) **104**, 575. (Chap. 1)

Rösler M., Brauer W. [1984]: Phys. Stat. Sol. (b) **126**, 629. (Chap. 1)

Rösler M., Brauer W. [1988]: Phys. Stat. Sol. (b) **148**, 213. (Chap. 1)

Schiff L.I. [1955]: *Quantum Mechanics* (Mc Graw-Hill Book Co., New York). (Sect. 2.1)

Schmid R., Gaukler K.H., Seiler H. [1983]: Scanning Electron Microscopy **2**, 501. (Sect. 5.1.2)

Schou J. [1980]: Phys. Rev. **B22**, 2141. (Chapts. 1, 5, Sect. 4.5)

Schou J. [1988]: Scanning Microscopy **2**, 607. (Chapts. 1, 5, Sect. 4.5)

Shimizu R., Everhart T.E. [1978]: Appl. Phys. Lett. **33**, 784. (Sect. 2.3.3)

Shimizu R., Ikuta T., Murata K. [1972]: J. Appl. Phys. **43**, 4233. (Sect. 2.1)

Sigmund P., Tougaard S. [1981]: "Electron emission from solids during ion bombardment. Theoretical aspects", in *Inelastic Particle-Surface Collisions*, ed. by Taglauer E., Heiland W. (Springer-Verlag, Berlin) pp. 2–37. (Chap. 1)

Smrcka L. [1970]: Czech. J. Phys. **B20**, 291. (Sects. 2.1, 5.1)

Snow E.C. [1967]: Phys. Rev. **158**, 683. (Sect. 2.1)

Spencer L.V. [1959]: **1**, N.B.S. Monograph (Washington). (Sect. 3.2)

Sternglass E.J. [1957]: Phys. Rev. **108**, 1. (Chap. 1)

Stolz H. [1959]: Ann. Phys. **7[3]**, 197. (Chap. 1)

Streitwolf H.W. [1959]: Ann. Phys. **7[3]**, 183. (Chap. 1)

Sturm K. [1982]: Adv. Phys. **31**, 1. (Sect. 2.2)

Svensson B., Holmen G., Buren A. [1981]: Phys. Rev. **B24**, 3749. (Sect. 5.4)

Tanuma S., Powell C.J., Penn D.R. [1988]: Surf. Interf. Anal. **11**, 577. (Sect. 2.2, 2.3.2)

Tholomier M., Doghmane N., Vicario E. [1988]: J. Microsc. Spectrosc. Electron. **13**, 119. (Sect. 2.1)

Tholomier M., Vicario E., Doghmane N. [1987]: J. Microsc. Spectrosc. Electron. **12**, 449. (Sect. 2.1)

Thomas S., Pattinson E.B. [1970]: J.Phys. **D3**, 349. (Sects. 5.1.3, 5.2, 5.3)

Tosatti E., Parravicini G.P. [1971]: J. Phys. Chem. Solids **32**, 623. (Sect. 2.3.1)

Tung C.J., Ashley J.C., Birkhoff R.D., Ritchie R.H., Emerson L.C., Anderson V.E. [1977]: Phys. Rev. **B16**, 3049. (Sect. 2.3.1)

Tung C.J., Ashley J.C., Ritchie R.H. [1979]: Surf. Sci. **81**, 427 . (Sect. 2.3.2)

Vashishta P., Singwi K.S. [1972]: Phys. Rev. **B6**, 875. (Sect. 2.2.1, 5.1.3)

Wehenkel C. [1975]: J. Physique **36**, 199. (Sect. 5.2)

Werner U., Heydenreich J. [1984]: Ultramicroscopy **15**, 17. (Sects. 2.1, 3.3)

Williams M.M.R. [1966]: *The Slowing-Down and Thermalization of Neutrons* (North-Holland Publishing Company, Amsterdam). (Chap. 4)

Williams M.M.R. [1971]: *Mathematical Methods in Particle Transport Theory* (Butterworths and Co., London). (Chap. 4)

Williams M.M.R. [1979]: Progress in Nuclear Energy **3**, 1. (Chap. 4)

Wolff P.A. [1954]: Phys. Rev. **95**, 56. (Chap. 1, Sect. 4.2)

Subject Index

Angular distribution   3, 47, 48, 50, 97, 110

Backscattering coefficient   3, 4, 110, 119
Bloch sum   30, 40, 77
Boltzmann equation   14–16, 43, 44, 92
Boundary condition   90, 93, 94, 97

Cascade maximum   47, 50
Charge state of the ion   35

Dielectric function   10–13, 75, 115

Efficiency of backscattered electrons   4, 52, 119
Elastic peak   109
Elastic scattering   17, 21, 53, 54, 71
— first Born approximation   71
— partial wave expansion method   71
— Rutherford cross section   71
Electron capture and loss   122
Electron current   97, 100, 106
Electron-electron interaction   8, 18, 36, 37
Electron flux   90, 93
Electron yield   3, 4, 47–49, 50, 51, 94, 110, 115, 118
— backward yield   55, 121
— forward yield   55, 121
Energy angular distribution   3, 10, 97
Energy distribution   3, 45, 47, 50, 106, 108, 117
Energy loss function   12, 13, 79
— optical   79
Escape process   8–10, 68
— boundary condition   93, 94, 97
— cone   9, 91
— depth   4, 15, 118
Excitation function   15, 23, 24
— by Auger processes   31–33, 40
— by decay of plasmons   26ff, 38, 68, 77
— of core electrons   30, 31, 38–40, 77, 79, 80
— of single conduction electrons   25, 26, 35ff

Forward-backward yield ratio   55, 121
Free-electron-gas model   8, 12, 75

Green's function   96, 102

Homogeneous Excitation   16, 98

Interband transition   13, 19, 21, 77

Mean free path   14
— elastic   21, 22, 72, 73
— ineleastic   22, 23, 79
Model potential   19–21
Monoenergetic Excitation   44–46
Monte Carlo simulation   81
— direct simulation scheme   82
— continous slowing-down scheme   83
— multiple collision scheme   84
— variance reduction   83, 89
— statistics   113
Multigroup method   102, 104

Optical loss function   79
Orthogonalized plane wave   30, 40, 78

Phase shift   17, 21, 73
Plasmon   12
— damping   13, 20, 76
— decay   13, 19, 26ff, 38, 76, 106, 107
— shoulder   47, 50, 76, 107
Potential barrier   67, 93
Primary electron
— transport   67, 70
— penetration depth   111

Radial distribution   112
Random phase approximation   12, 75
Range of the primary particle   4, 15

Scattering function   16ff
— elastic   17, 88, 95
— inelastic   17ff, 88, 95

Statistics 88
Stopping power 56, 57, 77, 83, 105, 117
Surface zone 77

Thin films 55
Transition probability 10ff
Transport equation 14–16, 43, 44, 89ff
Transport model
— infinite medium slowing down 68, 69
— age-diffusion 69, 94

— transport-albedo 69, 101
— S_N-multigroup 102

Valence band 80

Work function 8, 93

N. G. Chetaev
Theoretical Mechanics

Translated from the Russian by I. Aleksanova

1989. 407 pp. 190 figs. Hardcover DM 68,- ISBN 3-540-51379-5

This university-level textbook reflects the extensive teaching experience of
N. G. Chetaev, one of the most influential teachers of theoretical mechanics in
the Soviet Union. The mathematically rigorous presentation largely follows the
traditional approach, supplemented by material not covered in most other books
on the subject. To stimulate active learning numerous carefully selected exer-
cises are provided. Attention is drawn to historical pitfalls and errors that have
led to physical misconceptions.

D. Park
Classical Dynamics and Its Quantum Analogues

2nd enl. and updated ed. 1990. IX, 333 pp. 101 figs. Hardcover DM 78,-
ISBN 3-540-51398-1

(Originally published as Vol. 110 in the series Lecture Notes in Physics, 1979)

The primary purpose of this textbook is to introduce students to the principles
of classical dynamics of particles, rigid bodies, and continuous systems while
showing their relevance to subjects of contemporary interest. Two of these
subjects are quantum mechanics and general relativity. The book shows in many
examples the relations between quantum and classical mechanics and uses classical methods to derive most of the observational
tests of general relativity. A third area of current
interest is in nonlinear systems, and there are
discussions of instability and of the geometrical
methods used to study chaotic behaviour. In the
belief that it is most important at this stage of a
student's education to develop clear conceptual
understanding, the mathematics is for the most
part kept rather simple and traditional. In the belief
that a good education in physics involves learning
the history of the subject, this book devotes some
space to important transitions in dynamics: development of analytical methods in the 18th century
and the invention of quantum mechanics.

Springer-Verlag
Berlin
Heidelberg
New York
London
Paris
Tokyo
Hong Kong
Barcelona
Budapest

B. N. Zakhariev, A. A. Suzko

Direct and Inverse Problems

Potentials in Quantum Scattering

1990. XIII, 223 pp. 42 figs. Softcover DM 48,– ISBN 3-540-52484-3

This textbook can almost be viewed as a "how-to" manual for solving quantum inverse problems, that is, for deriving the potential from spectra and/or scattering data. The formal exposition of inverse methods is paralleled by a discussion of the direct problem.
In part differential and finite-difference equations are presented side by side. A variety of solution methods is presented. Their common features and (dis)advantages are analyzed. To foster a better understanding, the physical meaning of the mathematical quantities are discussed in detail.
Wave confinement in continuum bound states, resonance and collective tunneling, and the spectral and phase equivalence of various interactions are some of the physical problems covered.

A. G. Sitenko

Scattering Theory

1991. XI, 294 pp. 32 figs. (Springer Series in Nuclear and Particle Physics) Hardcover DM 88,– ISBN 3-540-51953-X

This mathematically rigorous introduction to nonrelativistic scattering theory addresses upper level undergraduates in physics. The relationship between the scattering matrix and physical observables is discussed in detail. Among the emphasized topics are the stationary formulation of the scattering problem, the inverse scattering problem, dispersion relations, three-particle bound states and their scattering, collisions of particles with spin and polarization phenomena.
The analytical properties of the scattering matrix are discussed. Problems are included to help the reader to gain some experience and more expertise in scattering theory.

Springer-Verlag
Berlin
Heidelberg
New York
London
Paris
Tokyo
Hong Kong
Barcelona
Budapest

Springer Tracts in Modern Physics

* denotes a volume which contains a Classified Index starting from Volume 36